Earth Science & Astronomy for the Logic Stage
Student Guide Table of Contents

Letter to the Student	5
Date Sheets	9
Astronomy Unit 1: Space	**13**
Vocabulary Sheet	*14*
Universe	*16*
Galaxies	*22*
Stars	*28*
Constellations	*34*
Zodiac	*40*
Astronomy Unit 2: Our Solar System	**45**
Vocabulary Sheet	*46*
Sun	*48*
Inner Planets-Mercury, Venus, and Mars	*54*
Earth and the Moon	*60*
Outer Planets-Jupiter and Saturn	*66*
Outer Planets-Uranus, Neptune, and Minor Members	*72*
Comets and Meteors	*76*
Astronomy Unit 3 Astronomers & Their Tools	**83**
Vocabulary Sheet	*84*
Astronomers	*86*
Looking Into Space	*92*
Exploring Space	*98*
Satellites	*104*
Earth Science Unit 4: Our Planet	**109**
Vocabulary Sheet	*110*
Inside Earth	*112*
Maps and Mapping	*118*
Rivers	*124*
Oceans	*130*
Glaciers	*136*
Natural Cycles	*142*
Biomes and Habitats	*148*

Earth Science Unit 5: Geology — 155

    Vocabulary Sheet — 156
    Continents — 158
    Volcanoes — 164
    Earthquakes — 170
    Mountains — 176
    Rocks — 182
    Ores and Gems — 188
    Erosion and Weathering — 194

Earth Science Unit 6: Weather — 201

    Vocabulary Sheet — 202
    Atmosphere — 204
    Climates — 210
    Weather — 216
    Clouds — 222
    Extreme Weather — 228
    Forecasting — 234

Appendix — 241

    Astronomy Memory Work — 243
    Earth Science Memory Work — 244
    Article on the 4 Types of Galaxies — 246
    Article on Ursa Major & Ursa Minor — 247
    Article on Freeze-Thaw Weathering — 251
    Activity Log — 252

Glossary — 257

Earth Science & Astronomy for the Logic Stage

Student Guide

> **THIS PRODUCT IS INTENDED FOR HOME USE ONLY**
>
> The images and all other content in this book are copyrighted material owned by Elemental Science, Inc. Please do not reproduce this content on e-mail lists or websites. If you have an eBook, you may print out as many copies as you need for use WITHIN YOUR IMMEDIATE FAMILY ONLY. Duplicating this book or printing the eBook so that the book can then be reused or resold is a violation of copyright.
>
> **Schools and co-ops:** You MAY NOT DUPLICATE OR PRINT any portion of this book for use in the classroom. Please contact us for licensing options at support@elementalscience.com.

Earth Science & Astronomy for the Logic Stage Student Guide

Second Edition (Third Printing, 2020)
Copyright @ Elemental Science, Inc.

ISBN# 978-1-935614-61-6

Printed in the USA for worldwide distribution

For more copies write to:
Elemental Science
PO Box 79
Niceville, FL 32588
support@elementalscience.com

Copyright Policy

All contents copyright © 2012, 2014, 2016, 2017, 2020 by Elemental Science. All rights reserved.

Limit of Liability and Disclaimer of Warranty: The publisher has used its best efforts in preparing this book, and the information provided herein is provided "as is." Elemental Science makes no representation or warranties with respect to the accuracy or completeness of the contents of this book and specifically disclaims any implied warranties of merchantability or fitness for any particular purpose and shall in no event be liable for any loss of profit or any other commercial damage, including but not limited to special, incidental, consequential, or other damages.

Trademarks: This book identifies product names and services known to be trademarks, registered trademarks, or service marks of their respective holders. They are used throughout this book in an editorial fashion only. In addition, terms suspected of being trademarks, registered trademarks, or service marks have been appropriately capitalized, although Elemental Science cannot attest to the accuracy of this information. Use of a term in this book should not be regarded as affecting the validity of any trademark, registered trademark, or service mark. Elemental Science is not associated with any product or vendor mentioned in this book.

Earth Science & Astronomy for the Logic Stage
Letter to the Student

Dear Student,

Welcome to your journey through earth science and astronomy. Earth science is the study of the Earth, while astronomy is the study of space. This year you will examine the various landforms on the Earth, the weather, our solar system, and what can be found in space. You will look at mountains, tornadoes, planets, and stars along your voyage. This guide is written to you, so enjoy your journey!

What this guide contains

First, this guide includes your date sheets and unit sheets. The unit sheets include your vocabulary words, weekly student assignment sheets, sketches, experiment sheets, and space for each of your writing assignments. After your unit sheets, you will find the appendix of this guide. In it you will find a list of all your memory work for the year, a glossary, and a place to record any additional activities you have done that pertain to earth science and astronomy.

Student Assignment Sheets Explained

The Student Assignment Sheets contain your weekly assignments for each week. Each of the student assignment sheets contain the following:

- ✓ **Weekly Topic & Experiment**

    Each week will revolve around a weekly topic to be studied. You will be assigned an experiment to complete that poses a question about the topic studied. Your student assignment sheets contain the list of materials you will need and the instructions to complete the experiments. This guide also includes an experiment sheet for you to fill out. Each of the experiments will have you use the scientific method.

    A Word about the Scientific Method

    The scientific method is a method for asking and answering scientific questions. This is done through observation and experimentation. The following steps are key to the scientific method:

    1. **Ask A Question** – The scientific method begins with asking a question about something you observe. Your questions must be about something you can measure. Good questions begin with how, what, when, who, which, why, or where.
    2. **Do Some Research** – You need to read about the topic from your question so that you can have background knowledge of the topic. This will keep you from repeating mistakes that have been made in the past.
    3. **Formulate a Hypothesis** – A hypothesis is an educated guess about the answer to your question. Your hypothesis must be easy to measure and answer the original question you asked.
    4. **Test with Experimentation** – Your experiment tests whether your hypothesis is true or false. It is important for your test to be fair. This means that you may need to run multiple tests. If you do, be sure to only change one factor at a time so that

you can determine which factor is causing the difference.

5. **Record and Analyze Observations or Results** – Once your experiment is complete, you will collect and measure all your data to see if your hypothesis is true or false. Scientists often find that their hypothesis was false. If this is the case, they will formulate a new hypothesis and begin the process again until they are able to answer their question.

6. **Draw a Conclusion** – Once you have analyzed your results, you can make a statement about them. This statement communicates your results to others.

Each of your experiment sheets will begin with a question and an introduction. The introduction will give you some background knowledge for the experiment. The experiment sheet also contains sections for the materials, a hypothesis, a procedure, an observation, and a conclusion. In the materials listed section you need to fill out what you used to complete the experiment. In the hypothesis section you need to predict what the answer to the question posed in the lab is. In the procedure section you need to write step by step what you did during your experiment so that someone else could read your report and replicate your experiment. In the observation section you need to write what you saw. Finally, in the conclusion section you need to write whether or not your hypothesis was correct and any additional information you have learned from the experiment. If your hypothesis was not correct, discuss why with your teacher and then include why your experiment did not work on your experiment sheet.

Vocabulary & Memory Work

Throughout the year you will be assigned vocabulary and memory work for each unit. Each week you will need to look up the word in your glossary and fill out the definitions on the Unit Vocabulary sheet found at the beginning of each unit in this guide. You may also want to make flash cards to help you work on memorizing these words. Each week you will also have a memory work selection to work on. Simply repeat this selection until you have it memorized, and then say the selection to your teacher. There is a complete listing of the memory work selections in the appendix of this guide.

Sketches

Each week you will be assigned a sketch to complete. Color the sketch and label it with the information given on the Student Assignment sheet. Be sure to give your sketch a title.

Writing Assignment

Each week you will be writing an outline and/or a narrative summary. The student assignment page will give you a reading assignment for the topic from your spine text, either the *Kingfisher Science Encyclopedia* or the *Usborne Science Encyclopedia*. After you have finished the assignment, discuss what you have read with your teacher. Your teacher will let you know whether to write an outline or a narrative summary from your spine text reading. Your teacher may also assign additional research reading out of the following books:

- *The DK Encyclopedia of Science* (DKEOS)
- *Astronomy DK Eyewitness Book* (DK Astro)
- *Exploring the Night Sky* (ENS)

Once you finish the additional reading, prepare a narrative summary about what you have learned from your reading. Your outlines should be two-level main topic style outlines and your narrative summaries should be two to four paragraphs in length, unless otherwise assigned by your teacher.

- **Dates to Be Entered**

    Each week dates of important discoveries within the topic and dates from the readings are given on the student assignment sheet. You will enter these dates onto one of four date sheets. The date sheets are divided into the four time periods laid out in *The Well-Trained Mind* by Susan Wise Bauer and Jessie Wise (Ancients, Medieval-Early Renaissance, Late Renaissance-Early Modern, Modern). These sheets are found in the ongoing projects section of this guide. You can choose to just write the dates and information on the sheet or if you want you can draw a time line in the space provided and enter your dates on that.

How to schedule this study

Earth Science & Astronomy for the Logic Stage is designed to take up to three hours per week. You, along with your teacher, can choose whether to complete the work over five days or over two days. Below are two options for scheduling to give you an idea of how you can schedule your week:

- ✓ A typical two day schedule
    - Day 1 – Define the vocabulary, do the experiment, complete the experiment page, and record the dates.
    - Day 2 – Read assigned pages and discuss together, prepare the science report or outline, and complete the sketch.
- ✓ A typical five day schedule
    - Day 1 – Do the experiment and complete the experiment page.
    - Day 2 – Record the dates and define the vocabulary.
    - Day 3 – Read assigned pages and discuss together and complete the sketch.
    - Day 4 – Prepare the science report or outline.
    - Day 5 – Complete one of the Want More activities from the teacher guide,

Final Thoughts

As the author and publisher of this curriculum, I encourage you to contact me with any questions or problems that you might have concerning *Earth Science & Astronomy for the Logic Stage* at support@elementalscience.com. I will be more than happy to answer them as soon as I am able. I hope that you will enjoy *Earth Science & Astronomy for the Logic Stage*!

Sincerely,

Paige Hudson, BS Biochemistry, Author

Ancients 5000 BC–400 AD

Medieval-Early Renaissance 400AD-1600 AD

Late Renaissance-Early Modern 1600 AD-1850 AD

Modern 1850 AD–Present

Modern 1850 AD–Present

Astronomy Unit 1
Space

Astronomy Unit 1: Space
Vocabulary Sheet

Define the following terms as they are assigned on your Student Assignment Sheet.

1. Universe – _____
   _____
   _____

2. Galaxy – _____
   _____
   _____

3. Cluster – _____
   _____
   _____

4. Supercluster – _____
   _____
   _____

5. Star – _____
   _____
   _____

6. Nebulae – _____
   _____
   _____

7. Black hole – _____
   _____
   _____

8. Constellation – _____
   _____
   _____

9. Planetarium – _____
_____
_____

10. Zodiac – _____
_____
_____

Student Assignment Sheet Astronomy Week 1
Universe

Experiment: Can I calculate the speed of light using a microwave?

Materials
- ✓ Large chocolate bar
- ✓ Large plate
- ✓ Microwave
- ✓ Ruler

> **⚠ CAUTION**
> *Melted chocolate is very hot and can burn, DO NOT touch!*

Procedure
1. Read the introduction to this experiment and answer the question.
2. Unwrap the chocolate bar and place it on the plate. Remove the turntable from your microwave and set the plate inside.
3. Begin by heating the chocolate bar for 1 minute; check to see if it has begun to melt. (*You are looking to see if there are two spots that have begun to melt. You DO NOT want the entire chocolate bar to melt.*) If the bar has not begun to melt, continue to heat it for 30-second intervals, checking each time to see if melting has begun.
4. Once melting has begun, carefully remove the plate from the microwave and measure the distance between the centers of the two melted points.
5. Record the distance on your experiment sheet and complete the calculations.
6. Draw conclusions and complete your experiment sheet.

Vocabulary & Memory Work
- ☐ Vocabulary: universe
- ☐ Memory Work – Work on memorizing the Types of Stars.

Sketch Assignment: Contents of the Universe
- ▦ Label the following: Earth, Our Solar System, Milky Way Galaxy, Cluster of Galaxies, The Universe

Writing Assignment
- ↪ Reading Assignment: *Kingfisher Science Encyclopedia* pg. 386-387 The Universe
- ↪ Additional Research Readings
  - 📖 Big Bang Theory: *KSE* pg. 388-389
  - 📖 The Universe: *USE* pg. 154-155

Dates to Enter
- 🕐 1929 – Edwin Hubble proves that the universe is expanding.
- 🕐 1965 – Scientists find heat waves in the universe that they believe are leftover from a vast explosion.
- 🕐 1992 – The satellite *Cosmic Background Explorer* traces background radiation and ripples in the universe that are thought to be leftover from the Big Bang.

Sketch Assignment Week 1

Student Guide Astronomy Unit 1: Space ~ Week 1 Universe

Experiment: Can I calculate the speed of light using a microwave?

Introduction

The universe is the vast expanse of space in which all things are found. It contains us, our planet, our solar system, our galaxy and at least 100 billion other galaxies. Astronomers have studied what they call the observable universe since Galileo invented the first telescope in 1609. As astronomers began to record their observations, they found the need to calculate distances, so that they could give approximate locations for the objects in the universe. Since the distances in the universe are so large, astronomers began to use light years to calculate them. A light year is the distance a wavelength of light will travel in one year, or about 5.88 trillion miles. In this experiment, you are going to try to calculate the speed of light, which scientists then multiply by the time in one year to calculate a light year.

Hypothesis

Can I calculate the speed of light using a microwave?   Yes   No

Materials

_____   _____

_____   _____

_____   _____

_____   _____

Procedure

_____

_____

_____

_____

_____

_____

_____

_____

Observations

Distance from the center of one melted chocolate spot to the center of the next melted chocolate spot	
in cm (centimeters)	
in m (meters)	

_____
_____
_____
_____
_____
_____

Results

Calculation #1: _____ x 2 = _____
              (distance measured in m)   (wavelength in m)

Calculation #2: _____ x 1,000,000 (Hz/MHz) = _____
             (Mega Hertz(MHz), from the back of the microwave)   (Cycles per second in Hz)

Calculation #3: _____ x _____ = _____
            (wavelength in m)   (Cycles per second in Hz)   (speed of light m/s)

Conclusion

_____
_____
_____
_____

Written Assignment Week 1

Discussion Questions

1. How can we see other galaxies?
2. How do astronomers measure distances in the universe?
3. What does everything in the universe do?

Written Assignment Week 1

Student Assignment Sheet Astronomy Week 2
Galaxies

Experiment: Identifying Galaxies
   Materials
   ✓ Galaxy cards (Your instructor will provide these cards.)
   Procedure
   1. Read the introduction to this experiment and answer the question.
   2. Read "Four Types of Galaxies" found on pg. 246 of the Appendix. Then, fill in the Galaxy Information Chart on your experiment sheet with the information.
   3. Next, look at each of the galaxy cards, determine which type of galaxy they are by using the information included, and then check your answers with your teacher.
   4. Write what you have learned in the conclusion section of your experiment sheet.

Vocabulary & Memory Work
   ☐ Vocabulary: galaxy, cluster, supercluster
   ☐ Memory Work – Continue to work on memorizing the Types of Stars.

Sketch Assignment: 4 Types of Galaxies
   Label the following: irregular galaxy, elliptical galaxy, spiral galaxy and barred spiral galaxy

Writing Assignment
   Reading Assignment: *Kingfisher Science Encyclopedia* pg. 390-391 Galaxies
   Additional Research Readings
      Galaxies: *USE* pg. 156-157
      Our Galaxy and Beyond: *DK Astro* pg. 62-63

Dates to Enter
   🕒 1784 – Charles Messier finds several blurry objects that he records as nebulae. These are later discovered to be galaxies.
   🕒 1845 – Lord Rosse draws the galaxy M51, without knowing what it is.
   🕒 1924 – Edwin Hubble presents the first evidence of other galaxies.

Sketch Assignment Week 2

Student Guide Astronomy Unit 1: Space ~ Week 2 Galaxies

Experiment: Identifying Galaxies

Introduction

All galaxies are massive celestial bodies made up of stars, gas and dust. They are held together by gravity and classified by their shape. There are four main types of galaxy shapes, spiral, elliptical, barred spiral, and irregular. In this experiment, you are going to read more about each type of galaxy shape and then identify several galaxies using what you have learned.

Conclusion

Galaxy Information Chart

Type of Galaxy				
Shape				
Presence of Bulges & Discs				
Presence of Gas & Dust				
Types of Stars				
Examples (from Galaxy Cards)				

Written Assignment Week 2

Discussion Questions

1. What is the name of the galaxy our solar system is in?
2. How are galaxies classified?
3. What are two things that astronomers believe help to determine the shape of a galaxy?
4. Describe the typical active galaxy.
5. Which cluster of galaxies does the Milky Way belong to?

Written Assignment Week 2

Student Assignment Sheet Astronomy Week 3
Stars

Experiment: Why do stars twinkle?
- Materials
  - ✓ Aluminum foil
  - ✓ Large piece of cardboard (large enough to fit under the glass bowl)
  - ✓ Glass bowl
  - ✓ Flashlight
  - ✓ Scissors
- Procedure
  1. Read the introduction to this experiment and answer the question.
  2. Cut out small pieces of aluminum foil, shape them into stars and glue them onto the cardboard.
  3. Fill up two thirds of the bowl with water and place it on top of the cardboard.
  4. Turn off the lights in the room you are in and shine your flashlight on the top of the bowl.
  5. Tap the side of the bowl and record what happens.
  6. Discuss with your teacher, draw conclusions, and finish your experiment sheet.

Vocabulary & Memory Work
- ☐ Vocabulary: star, nebulae, black hole
- ☐ Memory Work – Continue to work on memorizing the Types of Stars.

Sketch Assignment: Life Cycle of a Star
- Label the following: stellar nebulae, average star, red giant, planetary nebula, neutron star, the material condenses and begins to burn by converting hydrogen to helium, helium starts to convert to carbon and the star gets larger, the fuel in the core is used up and the outer layers are pushed out into space, the core shrinks to the size of a small planet

Writing Assignment
- Reading Assignment: *Kingfisher Science Encyclopedia* pg. 392-393 Stars
- Additional Research Readings:
  - Stars: *USE* pg. 158-159
  - The birth and death of stars: *DK Astro* pg. 60-61

Dates to Enter
- 1824-1910 – British astronomer William Huggins lives. He is the first to use spectroscopy in astronomy.
- 1906 – Ejnar Hertzsprung, a Danish astronomer, discovers that a star's temperature and luminosity are linked. He then arranges them into families, from hot and bright to cool and dim.

Sketch Assignment Week 3

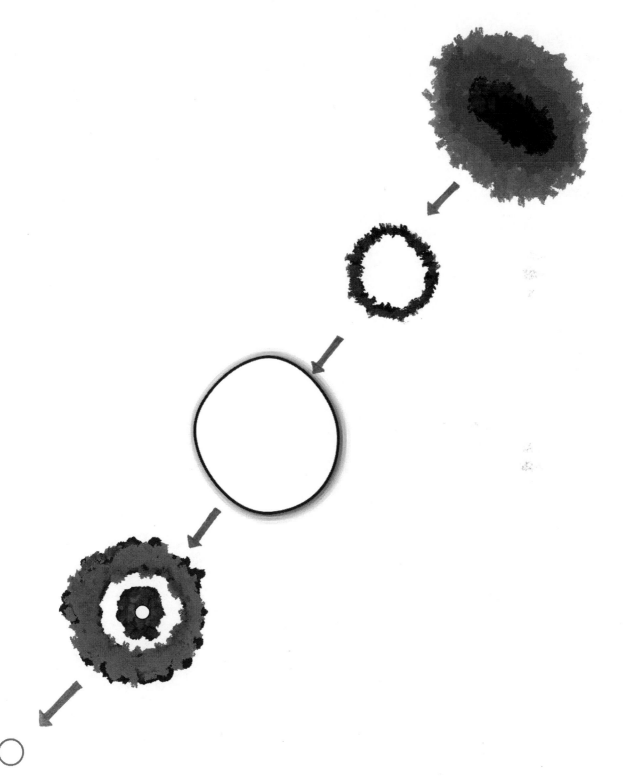

Student Guide Astronomy Unit 1: Space ~ Week 3 Stars

Experiment: Why do stars twinkle?

Introduction

Stars are the most numerous heavenly bodies in our universe. Our galaxy alone has about 200 billion of them. Stars are balls of hot, glowing gas that produce light that we can see from Earth. Stars vary in size, but all appear very small from the Earth because they are so far away. At times stars appear to sparkle in the night sky. In this experiment, you are going to figure out why they appear to twinkle.

Hypothesis

I think that stars twinkle because _____

_____

_____

_____

Materials

_____   _____

_____   _____

_____   _____

_____   _____

Procedure

_____

_____

_____

_____

_____

_____

_____

_____

Observations and Results

_____
_____
_____
_____
_____
_____
_____
_____

Conclusion

_____
_____
_____
_____
_____
_____
_____

Written Assignment Week 3

Discussion Questions

1. What does light coming from a star tell an astronomer about a star?
2. Name three types of energy that are produced by a star.
3. How does a star's mass affect it?
4. Name two characteristics that are common to all stars.
5. What are the two types of star clusters? Explain both.
6. What happens to the core of a star when it dies?

Written Assignment Week 3

Student Assignment Sheet Astronomy Week 4
The Big and Little Dipper

Experiment: Shoebox Planetarium
Materials
- ✓ Shoebox
- ✓ Black construction paper
- ✓ Pin
- ✓ Flashlight

Procedure
1. Read the introduction on your experiment sheet.
2. At one end of the shoebox, cut out a hole that is a little less than 3 inches wide and 5 inches long.
3. Cut out two rectangles from the black construction paper measuring 3 inches by 5 inches. Then using the templates from your experiment sheet and a pin, punch out the necessary holes to make the Big Dipper and the Little Dipper constellations.
4. Tape one of the cards in front of the hole that was cut in the shoebox earlier. Place the flashlight in the box and cover with the lid. Turn off the lights in the room, angle the shoebox towards a wall, and observe what you see.
5. Repeat with the second card.

Vocabulary & Memory Work
- ☐ Vocabulary: constellation, planetarium
- ☐ Memory Work – Begin to work on memorizing the Constellations of the Zodiac: Aquarius, Aries, Cancer, Capricorn, Gemini, Leo, Libra, Pisces, Sagittarius, Scorpio, Taurus, Virgo

Sketch Assignment: Big Dipper & Little Dipper
- Label the following: Ursa Major, Ursa Minor, Polaris, Dubhe, Merak, Mizar, Benetnasch

Writing Assignment
- Reading Assignment: "Ursa Major and Ursa Minor," "Callisto and Arcas," and "The Two Bears" from *The Stars and their Stories* (found in the Appendix of the SG pp. 115-117)
- Additional Research Readings
    - None this week

Dates to Enter
- 2nd century – Ptolemy names forty-eight different constellations in his book *Almagest*.
- 17th-18th centuries – Forty more constellations are named, for a total of eighty-eight named constellations.

Sketch Assignment Week 4

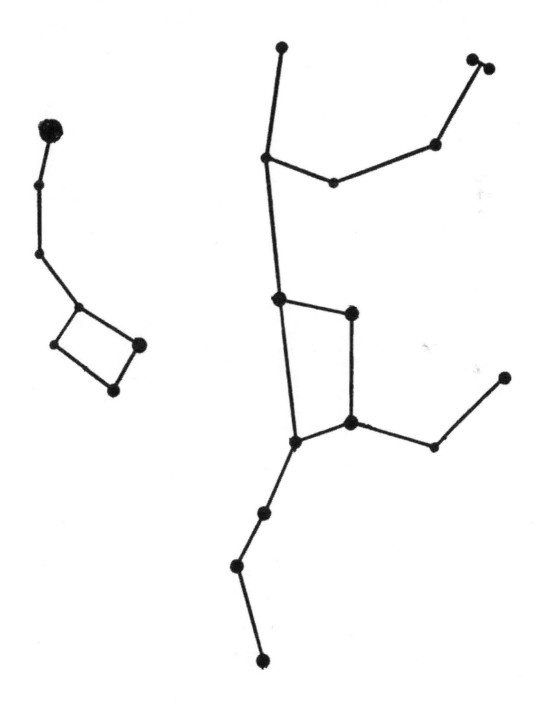

Student Guide Astronomy Unit 1: Space ~ Week 4 The Big and Little Dipper

Experiment: Shoebox Planetarium

Introduction

A constellation is a pattern or shape that the group of stars makes in the sky. There are 88 named constellations in our sky. Your position on the Earth and the time of year determine which constellations you can see. The most recognized constellations in the Northern Hemisphere are the Big Dipper and the Little Dipper. In this experiment, you are going to make your own planetarium, or device used to project images of the stars, so that you can view these constellations at anytime.

Templates for the Planetarium Cards

The Big Dipper Template

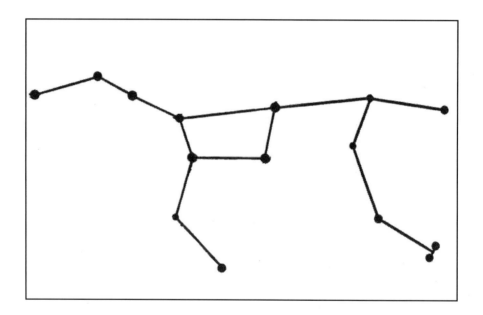

The Little Dipper Template

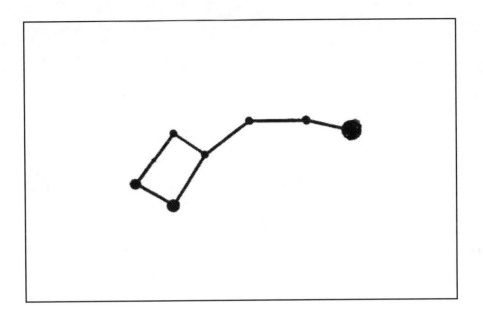

Written Assignment Week 4

Discussion Questions

1. Why are the constellations Ursa Major and Ursa Minor called the Big Dipper and Little Dipper?
2. Tell the story of the two bears in your own words.
3. What are some other names for the constellation Ursa Major?

Written Assignment Week 4

Student Assignment Sheet Astronomy Week 5
Constellations

Experiment: Constellations Research Project

This week you will spend part of the week on researching a constellation of the Zodiac and creating a profile page for that constellation.

Steps to Follow
1. Choose a Constellation – Choose one of the constellations of the Zodiac for an in-depth profile.
2. Do some research about the constellation – Use the internet and the resources you have in your home or at your library to find out more about your chosen constellation.
    Answer the following questions from your research:
    - ✓ What is the English translation of the Latin name of the constellation?
    - ✓ Where is the constellation found?
    - ✓ What are the major stars found in the constellation?
    - ✓ What is the best season to view the constellation?
    - ✓ What is the story behind the constellation's name?
    - ✓ Write down any interesting facts you have learned on individual index cards.
3. Complete the profile page for your constellation – Write the Latin name for your constellation on the blank at the top of the page. Then, fill out the remaining information. Your summary of the story behind the constellation should be four to five sentences long. Finally, draw your constellation in the box.

Writing Assignment
- Reading Assignment: *Kingfisher Science Encyclopedia* pg. 396-397 Constellations
- Additional Research Readings:
    - Constellations: *USE* pg. 160
    - Astrology: *DK Astro* pg. 16-17

Vocabulary & Memory Work
- ☐ Vocabulary: zodiac
- ☐ Memory Work – Continue to work on memorizing the Constellations of the Zodiac

Sketch Assignment: Constellations
- Label the following constellations and tell whether the are found in the Northern or Southern Hemisphere: The Southern Cross, Libra, Scorpio, Leo, Pegasus

Dates to Enter
- 🕓 7th century BC – Babylonian astronomers use a coordinate system resembling the Zodiac.
- 🕓 ca. 50 BC – A relief called the Dendera zodiac is the first known depiction of the classical zodiac of twelve signs.

Sketch Assignment Week 5

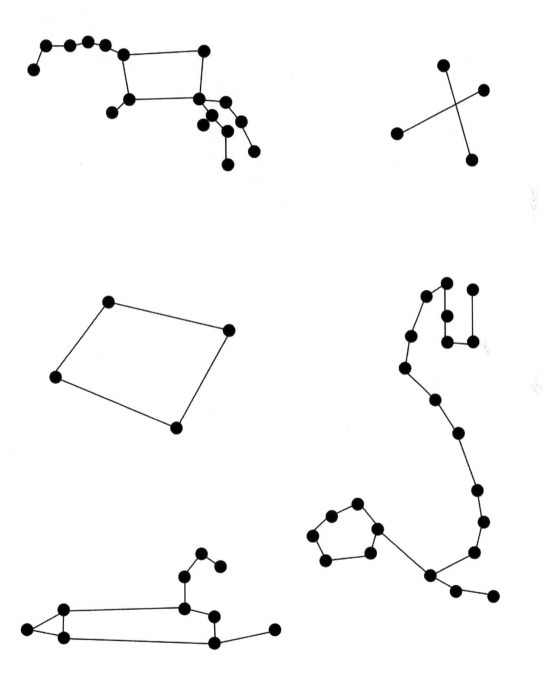

Student Guide Astronomy Unit 1: Space ~ Week 4 The Big and Little Dipper

Constellation Profile Page

English Translation _____

Location _____

_____

Major Stars _____

_____

_____

_____

Best season to view the constellation

_____

Story Behind the Name

_____
_____
_____
_____
_____
_____
_____
_____
_____
_____
_____
_____
_____

Astronomy Unit 2
Our Solar System

Astronomy Unit 2: Our Solar System
Vocabulary Sheet

Define the following terms as they are assigned on your Student Assignment Sheet.

1. Photosphere – _____
   _____
   _____

2. Prominence – _____
   _____
   _____

3. Solar Wind – _____
   _____
   _____

4. Sunspot – _____
   _____
   _____

5. Crater – _____
   _____
   _____

6. Planet – _____
   _____
   _____

7. Axis – _____
   _____
   _____

8. Orbit – _____
   _____
   _____

9. Moon – _____
_____
_____

10. Eclipse – _____
_____
_____

11. Galilean moons – _____
_____
_____

12. Gas giant – _____
_____
_____

13. Dwarf planet – _____
_____
_____

14. Asteroid – _____
_____
_____

15. Comet – _____
_____
_____

16. Meteor – _____
_____
_____

17. Meteoroid – _____
_____
_____

Student Assignment Sheet Astronomy Week 6
Sun

Experiment: Do the sun's rays contain heat?

Materials
- ✓ 2 glass jars or 2 clear glasses
- ✓ Plastic wrap
- ✓ 2 black tea bags
- ✓ Water
- ✓ Instant read thermometer

Procedure

NOTE – You will need to do this experiment on a sunny day.****
1. Read the introduction to this experiment and answer the question.
2. Fill each jar with the same amount of water. Take the temperature of the water in each jar. Then, place a tea bag in each jar and cover with the plastic wrap.
3. Set one of the jars outside in a sunny spot and leave the other jar inside your house.
4. After two hours, bring the jar you left outside into the house. Then, measure the temperature of the water and observe any changes to the appearance of the water. Write your results on your experiment sheet.
5. Draw conclusions and complete your experiment sheet.

Vocabulary & Memory Work
- ☐ Vocabulary: photosphere, prominence, solar wind, sunspot
- ☐ Memory Work – Begin to work on memorizing the Planet Order.
    1. Sun
    2. Mercury (Gravity: 0.38)
    3. Venus (Gravity: 0.90)
    4. Earth (Gravity: 1 or 9.8 m/s2) [Moon (Gravity: 0.17)]
    5. Mars (Gravity: 0.38)
    6. Jupiter (Gravity: 2.34)
    7. Saturn (Gravity: 0.93)
    8. Uranus (Gravity: 0.90)
    9. Neptune (Gravity: 1.13)

Sketch Assignment: Parts of the Sun
- Label the Following: chromosphere, flare, facula, prominence, sunspot, hydrogen core, photosphere

Writing Assignment
- Reading Assignment: *Kingfisher Science Encyclopedia* pp. 394-395 The Sun
- Additional Research Readings
    - The Sun: *USE* pp. 192-193
    - The Sun: *DK Astro* pp. 38-39

Dates to Enter:
- 1930 – French astronomer, Bernard Lyot, invents the coronagraph, allowing scientists to view the Sun without waiting for a total solar eclipse.
- 1868-1938 – George Hale lived. He was an American astronomer who studied sunspots and discovered the magnetic fields within them.

Sketch Assignment Week 6

Student Guide Astronomy Unit 2: Our Solar System ~ Week 6: Sun

Experiment: Do the sun's rays contain heat?

Introduction

The sun is the closest star to the earth. It is a giant ball of a constantly exploding gas called hydrogen. The surface of the sun is around 9,932°F, but the core is much hotter. It is the largest object in our Solar System. The sun's gravitational pull is so strong that all the planets in our Solar System revolve around it. The sun is important to the earth in many ways. In this experiment, you are going to look specifically at the sun's rays affect the objects on our planet.

Hypothesis

I think that the Sun's rays (do   /   do not   ) contain heat.

Materials

_____   _____
_____   _____
_____   _____
_____   _____
_____   _____

Procedure

_____
_____
_____
_____
_____
_____
_____
_____
_____
_____
_____
_____

Observations and Results

Temperature of the Jars

	Initial Temperature	Final Temperature
Outside Jar		
Inside Jar		

*Be sure to include units for your temperature measurements.

Appearance of the Jars

Outside Jar	Inside Jar

_____
_____
_____
_____
_____

Conclusion

_____
_____
_____
_____
_____

Written Assignment Week 6

Discussion Questions

1. What is the sun?
2. What causes sunspots?
3. What is the chromosphere of the sun composed of?
4. Briefly explain the expected life of the sun.

Written Assignment Week 6

Student Assignment Sheet Astronomy Week 7
Inner Planets–Mercury, Venus, and Mars

Experiment: Can yeast survive on Mercury or Venus?

Materials
- 3 heavy duty plastic bottles
- 3 balloons
- 3 tsp of yeast (1 ½ packages)
- 3 TBSP of sugar
- Water
- White vinegar

> ☹ **CAUTION**
> *Hot water is very dangerous and can cause injury. Be sure to have an adult handle the hot water at all times. ALWAYS use proper protection when handling hot water!*

Procedure
1. Read the introduction to this experiment and answer the question.
2. Label the bottles, "Mercury," "Venus," and "Earth." Use a funnel to carefully add 1 ½ cups of warm (between 125°F and 130°F) water to the Earth bottle. Use a funnel to carefully add 1½ cups of ice cold (below 40°F) water to the Mercury bottle. Have an adult use a funnel to carefully add 1 cup of hot (above 140°F) water and ½ cup of vinegar to the Venus bottle.
3. Add 1 tsp of yeast and one TBSP of sugar to each of the bottles, and then quickly cover the top of the bottle with a balloon.
4. Wait 30 minutes. Then, make observations and measure the circumference of each balloon.
5. Draw conclusions and complete your experiment sheet.

Vocabulary & Memory Work
- [ ] Vocabulary: crater, planet
- [ ] Memory Work – Continue to work on memorizing on Planet Order

Sketch Assignment: Inner Planets
- Label the Following: Mercury, Caloris Basin, crater-covered surface; Venus, multiple layers of dense clouds; Mars, Polar regions covered with water, ice and carbon dioxide ice

Writing Assignment
- Reading Assignment: *Kingfisher Science Encyclopedia* pg. 403 Mercury & pg. 404 Venus, pg. 405 Mars
- Additional Research Readings
  - The Inner Planets: *USE* pg. 164-165
  - Mercury: *DK Astro* pg. 44-45, Venus: *DK Astro* pg. 46-47, Mars: *DK Astro* pg. 48-49
  - Venus & Mercury: *ENS* pg. 28-29, *ENS* pg. 30-31

Dates to Enter
- 1974-1975 – *Mariner 10* flies past Mercury three times to take overlapping photos of the surface of the planet.
- 1990-1994 – *Magellan* maps 98% of the surface of Venus.
- 1976 – Two American space probes land on Mars to test for signs of life, but find none.
- 1997 – *Pathfinder* lands on Mars and delivers the robotic rover *Sojourner* to explore the surface of Mars.
- 2004 – Mobile robots, called *Spirit* and *Opportunity,* explore the surface of Mars.

Sketch Assignment Week 7

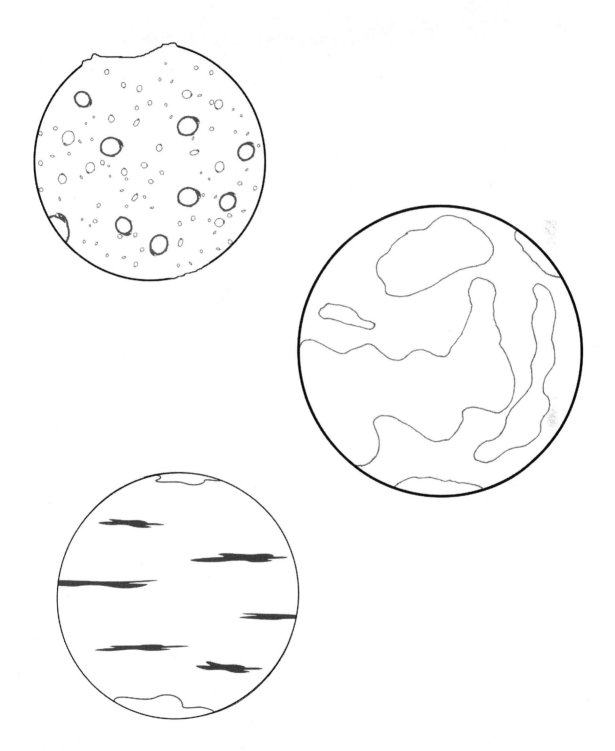

Student Guide Astronomy Unit 2: Our Solar System ~ Week 7: Inner Planets (Mercury, Venus, Mars)

Experiment: Can yeast survive on Mercury or Venus?

Introduction

Each planet has its own unique atmosphere or layer of gas surrounding the planet. The atmosphere and the planet's proximity to the Sun affect the surface conditions and temperature of the sphere. The atmosphere around Earth is perfectly suited to support life on the planet. In this experiment, you are going to test and see if Mercury and Venus could also support life.

Hypothesis

Can life survive on Mercury?   yes   no

Can life survive on Venus?   yes   no

Materials

Procedure

Observations and Results

Bottle	Circumference of the Balloon
Mercury	
Venus	
Earth	

Conclusion

Written Assignment Week 7

Discussion Questions

Mercury pg. 403
1. What is the surface of Mercury like?
2. What two factors contribute to the fact that a day on Mercury is longer than a year?

Venus pg. 404
1. Why does Venus appear as a bright star in our sky?
2. What are two reasons why humans could not survive on Venus?
3. What is the surface of Venus like?

Mars pg. 405
1. Why is Mars known as the Red Planet?
2. Has life been found on Mars? What have explorers found?
3. What is the surface of Mars like?

Written Assignment Week 7

Student Assignment Sheet Astronomy Week 8
Earth and the Moon

Experiment: Does the moon change size?

Materials
- ✓ Apple
- ✓ Fork
- ✓ Flashlight
- ✓ Partner

Procedure
1. Read the introduction to this experiment and answer the question.
2. Secure the apple on the end of the fork.
3. Have your partner stand 5 feet away from you and shine the flashlight in your direction.
4. Hold the fork with the apple at arm's length in front of the flashlight and a little above the beam. *(Note – The apple should appear darkened.)*
5. Keep the apple in the same place and turn slowly counterclockwise around an imaginary circle with the apple as the center. Continue until you have completed your circle and are back to your starting point. *(Note – You may need to duck down a bit as you turn so that you don't block the flashlight beam with your body.)*
6. Write down your observations about what happened to your apple as you were turning.
7. Draw conclusions and complete your experiment sheet.

Vocabulary & Memory Work
- ☐ Vocabulary: axis, orbit, moon, eclipse
- ☐ Memory Work – Continue to work on memorizing the Planet Order

Sketch Assignment: Phases of the Moon
- Label the following and color in the portion of the moon that the labels represent: New Moon, Waxing Crescent, First Quarter, Gibbous Waxing, Full Moon, Gibbous Waning, Last Quarter, and Waning Crescent

Writing Assignment
- Reading Assignment: *Kingfisher Science Encyclopedia* pg. 400-401 Earth and the Moon, pg. 402 Eclipses
- Additional Research Readings
  - The Earth and the Moon: *USE* pg. 166-167
  - Earth: *DK Astro* pg. 42-43
  - Moon: *DK Astro* pg. 40-41

Dates to Enter
- 1647 – Johannes Hevelius publishes the first lunar atlas.
- October 1959 – Russian spacecraft *Luna 3* transmits the first images of the far side of the moon back to the Earth.
- July 20, 1969 – US astronauts Neil Armstrong and Edwin Aldrin became the first humans to walk on the moon.

Sketch Assignment Week 8

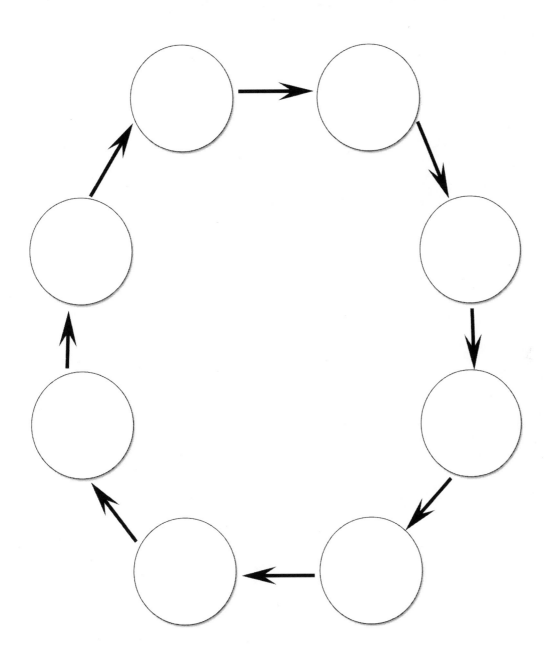

Student Guide Astronomy Unit 2: Our Solar System ~ Week 8: Earth/Moon

Experiment: Does the moon change size?

Introduction

　　The moon is a ball of rock that orbits around the Earth. It takes 27.3 days to complete its orbit around our planet. As the moon travels around the Earth it appears to be smaller or larger to us. In this experiment, you are going to determine if the moon actually changes size as it orbits around us by using an apple for the moon, a flashlight for the sun and your head for the Earth.

Hypothesis

Does the moon change size?　　　　yes　　　　　　　　no

Materials

_____　　_____
_____　　_____
_____　　_____
_____　　_____

Procedure

_____
_____
_____
_____
_____
_____
_____
_____
_____
_____

Observations and Results

Conclusion

Written Assignment Week 8

Discussion Questions

Earth and the Moon pg. 400-401
1. What makes the Earth unique to our solar system?
2. How long does it take for the Earth to orbit the sun one time? How long does it take the Earth to complete one spin on its axis?
3. Why do we only see one side of the moon from the Earth?
4. What is a moon phase?

Eclipses pg. 402
1. What is the difference between a solar eclipse and a lunar eclipse?

Written Assignment Week 8

Student Assignment Sheet Astronomy Week 9
Outer Planets–Jupiter and Saturn

Experiment: Does the heat in Jupiter's core affect the storms on its surface?

Materials
- ✓ Wooden dowel rod
- ✓ Paper clip
- ✓ 4 inch square of aluminum foil (not heavy duty)
- ✓ Toaster

> **CAUTION:** *Be careful not to touch your toaster or your foil kite until they have time to cool. They will be very hot and can burn you.*

Procedure
1. Read the introduction to this experiment and answer the question.
2. Begin by making a kite. First, unbend your paper clip halfway. Place the remaining loop of the paper clip over your wooden dowel (it should be a tight fit.) Then, puncture the corner of your aluminum foil with the straight end of the paper clip. Finally, bend the straight end of the paper clip over so that the foil is secure.
3. Turn your toaster on and wait 1 minute. Move your foil kite so that it is about a foot above the toaster's opening. Observe what happens to your kite and write it down on your experiment sheet.
4. Remove your kite from over the heat and turn off the toaster. Observe what happens to your kite and write it down on your experiment sheet.
5. Draw conclusions and complete your experiment sheet.

Vocabulary & Memory Work
- ☐ Vocabulary: Galilean moons, gas giants
- ☐ Memory Work – Continue to work on memorizing the Planet Order

Sketch Assignment: Jupiter & Saturn
- Label the following: Jupiter, the Red Spot, bands of storms, Saturn, rings, bands of ammonia, water and methane clouds

Writing Assignment
- Reading Assignment: *Kingfisher Science Encyclopedia* pg. 406 Jupiter, pg. 407 Saturn
- Additional Research Readings
    - The Outer Planets: *USE* pg. 168-169
    - Jupiter: *DK Astro* pg. 50-51, *ENS* pg. 32-33
    - Saturn: *DK Astro* pg. 52-53, *ENS* pg. 34-35

Dates to Enter
- 1610 – Galileo makes the first systematic study of Jupiter's 4 largest moons.
- 1660 – Robert Hooke reports a giant spot on Jupiter's surface.
- 1655 – Dutch scientist, Christiaan Huygens, correctly identifies Saturn's rings.
- 1979 – *Pioneer II* explores Saturn's rings.
- 1995 – The *Galileo* probe reaches Jupiter. It studies Jupiter's atmosphere and its moons.
- 2004 – *Cassini* arrives at Saturn to study the planet's moons and rings.

Sketch Assignment Week 9

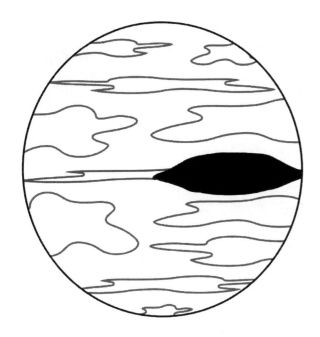

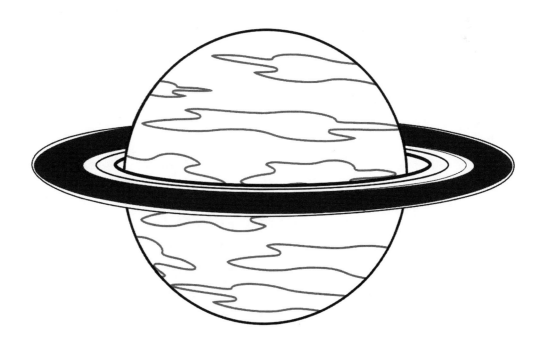

Student Guide Astronomy Unit 2: Our Solar System ~ Week 9: Outer Planets (Jupiter, Saturn)

Experiment: Does the heat in Jupiter's core affect the storms on its surface?

Introduction

Jupiter is a very turbulent planet. Its many bands and prominent red spot are evidence of a stormy atmosphere that surrounds the planet. Jupiter's Red Spot is known as the solar system's oldest hurricane. The high pressure that is found on the planet contributes to the perpetuation of these storms, but there are other factors involved as well. In this experiment, you will examine whether the heat from Jupiter's core is able to create winds that help to perpetuate the storms that cover the planet.

Hypothesis

I think that the heat in Jupiter's core (does / does not) affect the storms on its surface.

Materials

_____   _____

_____   _____

_____   _____

_____   _____

_____   _____

Procedure

_____

_____

_____

_____

_____

_____

_____

_____

Student Guide Astronomy Unit 2: Our Solar System ~ Week 9: Outer Planets (Jupiter, Saturn)

Observations and Results

Conclusion

Written Assignment Week 9

Discussion Questions

Jupiter pg. 406
1. Describe what Jupiter is like.
2. What are two factors that contribute to Jupiter's stormy atmosphere?

Saturn pg. 407
1. What is the outer surface of Saturn like?
2. What are Saturn's rings made from? Where are the largest pieces of Saturn's rings found?

Written Assignment Week 9

Student Assignment Sheet Astronomy Week 10
Outer Planets–Uranus, Neptune, and Minor Members

Experiment: Scaled Model of the Solar System

Materials
- ✓ Construction paper
- ✓ Paints or crayons

Procedure
1. This week you will be making a scaled version of the solar system, using the chart on your experiment sheet for your placements. Try to make your planets look as realistic as possible by painting or drawing the major features onto your models. For the Sun, tape four pieces of construction paper together lengthwise and then trim them up to look like a portion of a circle. Add solar flares, a prominence, sunspots, and facula.

	Distance from the Sun	Scale diameter of planet
Mercury	2 inches	3/4 inch
Venus	3 inches	1 3/4 inches
Earth	4 inches	2 inches
Mars	6 inches	1 1/8 inches
Jupiter	1 foot, 9 inches	22 inches
Saturn	3 feet, 2 inches	20 inches
Uranus	6 feet, 5 inches	8 inches
Neptune	10 feet, 1 inch	7 1/2 inches
Pluto (Optional)	13 feet, 3 inches	1/2 inches

2. Take a picture of your model and write what you have learned about the solar system from making the model on your experiment sheet.

Vocabulary & Memory Work
- ☐ Vocabulary: dwarf planet, asteroid
- ☐ Memory Work – Continue to work on memorizing the Planet Order

Writing Assignment
- ᨖ Reading Assignment: *Kingfisher Science Encyclopedia* pg. 408 Uranus, pg. 409 Neptune and pg. 410-411 The Solar System's Minor Members
- ᨖ Additional Research Readings:
  - 📖 The Outer Planets: *USE* pg. 170-171
  - 📖 Uranus: *DK Astro* pg. 54-55
  - 📖 Neptune & Pluto: *DK Astro* pg. 56-57

Dates to Enter
- 🕒 March 13, 1781 – William Herschel discovers Uranus using a homemade telescope.
- 🕒 1846 – Johann Galle finds the planet Neptune.
- 🕒 1930 – Clyde Tombaugh, a US astronomer, discovers Pluto.
- 🕒 2006 – Pluto is reclassified as a dwarf planet.

Scaled Model of the Solar System

What I learned from this project

_____
_____
_____
_____
_____
_____
_____
_____
_____
_____

Written Assignment Week 10

Discussion Questions

Uranus pg. 408
1. What are some key features of Uranus?

Neptune pg. 409
1. What are some key features of Neptune?
2. How was Neptune discovered?

The Solar System's Minor Members pg. 410-411
1. What are the four main groups of minor members in our solar system?
2. Why does Pluto's distance from the Sun vary?
3. Why are the asteroids in the asteroid belt called the minor planets?

Written Assignment Week 10

Student Assignment Sheet Astronomy Week 11
Comets and Meteors

Experiment: Does the size of a meteor change the impact it will have on a planet's surface?

Materials
- ✓ Marble
- ✓ Rubber bouncy ball
- ✓ Tennis ball
- ✓ Cake pan
- ✓ Cornstarch
- ✓ Ruler
- ✓ Foam ball (a little larger than the tennis ball)

Procedure
1. Read the introduction to this experiment and answer the question.
2. Pour a layer of cornstarch on the bottom of the cake pan about ½ inch deep. Shake lightly so that the surface is smooth.
3. Drop the marble from a height of 2 feet, aiming for the center of the pan. Observe what happens. Remove the marble, being careful not disturb the cornstarch, and measure the width and depth of the crater created. Record the measurement on your experiment sheet and shake the pan lightly so that the surface is smooth.
4. Repeat step #3 for the rubber bouncy ball, foam ball, and tennis ball.
5. Draw conclusions and complete your experiment sheet.

Vocabulary & Memory Work
- ☐ Vocabulary: comet, meteor, meteoroid
- ☐ Memory Work – Continue to work on memorizing the Planet Order

Sketch Assignment: Anatomy of a Comet
- ☒ Label the following: nucleus, dust tail, gas tail

Writing Assignment
- ෴ Reading Assignment: *Kingfisher Science Encyclopedia* pg. 412 Comets, pg. 413 Meteors and Meteorites
- ෴ Additional Research Readings
  - 📖 Space Debris: *USE* pg. 172-173
  - 📖 Travelers in Space: *DK Astro* pg. 58-59

Dates to Enter
- 🕐 1997 – The comet *Hale-Bopp* is in the Earth's night sky. It won't return for another 2,400 years.
- 🕐 1656-1742 – English astronomer, Edmond Halley lives. He correctly predicts that a comet will return to Earth's night sky in 1758, 1835, and 1910.
- 🕐 1976 – The second largest iron & nickel meteorite in the United States is found.
- 🕐 1986 – The space probe *Giotto*, is sent inside Halley's comet, giving people the first look at a comet's nucleus.

Sketch Assignment Week 11

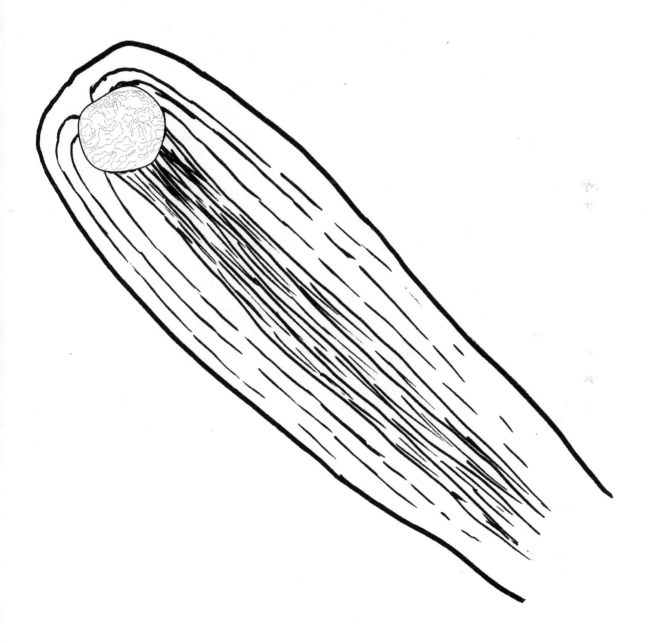

Student Guide Astronomy Unit 2: Our Solar System ~ Week 11: Comets and Meteorites

Experiment: Does the size of a meteor change the impact it will have on a planet's surface?

Introduction:

    Bits of space dust and iron-rich rock, called meteoroids, can be found throughout the solar system. They are constantly bombarding the planets. On Earth, most of these meteors burn up before they hit the surface due to our atmosphere, but this is not always the case on other planets. The surfaces of those planets are dotted with impact craters from the meteors. In this experiment, you are going to test whether the size of the meteor affects the size of the crater it will make.

Hypothesis (What I think the answer to the question will be):

    I think that the (larger / smaller) a meteor is the (larger / smaller) the impact crater will be.

Materials (What we used):

_____   _____

_____   _____

_____   _____

_____   _____

Procedure (What we did):

_____

_____

_____

_____

_____

_____

_____

_____

_____

Observations & Results (What I saw):

	Width of Crater	Depth of Crater
Marble		
Rubber Bouncy Ball		
Tennis Ball		
Foam Ball		

*Be sure to include units for your measurements.

Conclusion (What I learned):

Written Assignment Week 11

Discussion Questions
Comet pg. 412
1. What is common to all comets?
2. Why do comets grow a tail when they are closer to the sun?

Meteors & Meteorites pg. 413
1. What is a shooting star?
2. What are meteorites?

Written Assignment Week 11

Astronomy Unit 3

Astronomers & Their Tools

Astronomy Unit 3: Astronomers & Their Tools
Vocabulary Sheet

Define the following terms as they are assigned on your Student Assignment Sheet.

1. Astronomer –

2. Telescope –

3. Radio Telescope –

4. Reflecting Telescope –

5. Refracting Telescope –

6. Space Probe –

7. Rocket –

8. Space Shuttle –

9. Natural Satellite – _____
   _____
   _____

10. Artificial Satellite – _____
    _____
    _____

Student Assignment Sheet Astronomy Week 12
Astronomers

Experiment: Science Fair Project

This week, you will complete step one and begin step two of your Science Fair Project. You will be choosing your topic, formulating a question, and doing some research about that topic.

1. **Choose your topic** – You should choose a topic in the field of earth science or astronomy that interests you, such as comets. Next, come up with several questions you have relating to that topic, (e.g. "How fast does a comet melt?" or "How does the tail of a comet form?"). Then, choose the one question you would like to answer and refine it (e.g. "Does the size of a comet affect how fast it melts?").

2. **Do Some Research** – Now that you have a topic and a question for your project, it is time to learn more about your topic so that you can make an educated guess (hypothesis) on the answer to your question. For the question stated above, you would need to research topics like comets and how their tails form. Begin by looking up the topic in the references you have at home. Then, make a trip to the library to search for more on the topic. As you do your research, write any relevant facts you have learned on index cards and be sure to record the sources you use.

Vocabulary & Memory Work

- Vocabulary: astronomer, telescope
- Memory Work – Begin to work on memorizing the Ten Nearest Galaxies and Their Types.

  1. Milky Way (spiral)
  2. Sagittarius (elliptical)
  3. Large Magellanic Cloud (irregular)
  4. Small Magellanic Cloud (irregular)
  5. Ursa Minor (elliptical)
  6. Draco (elliptical)
  7. Sculptor (elliptical)
  8. Carina (elliptical)
  9. Sextans (elliptical)
  10. Fornax (elliptical)

Sketch Assignment: Astronomers Through the Ages

- Label the following:
  - → Nicolaus Copernicus (1473-1543) – He said that the Earth revolved around the Sun.
  - → Galileo Galilei (1564-1642) – He used the telescope to prove that the Earth and other planets move around the Sun.
  - → Isaac Newton (1643-1727) – He built the first reflective telescope and showed that the gravity we have on Earth is also in the universe.
  - → William Herschel (1738-1822) – He made a detailed catalog of nebulas and clusters, as well as discovered Uranus.
  - → Fred Hoyle (1915-2001) – He showed that carbon and oxygen are created in stars.

Writing Assignment

- Reading Assignment: *Kingfisher Science Encyclopedia* pg. 414-415 Astronomers
- Additional Research Readings
  - Copernican Revolution: *DK Astro* pg. 18-19
  - The Astronomer: *DK Astro* pg. 28-29

Sketch Assignment Week 12

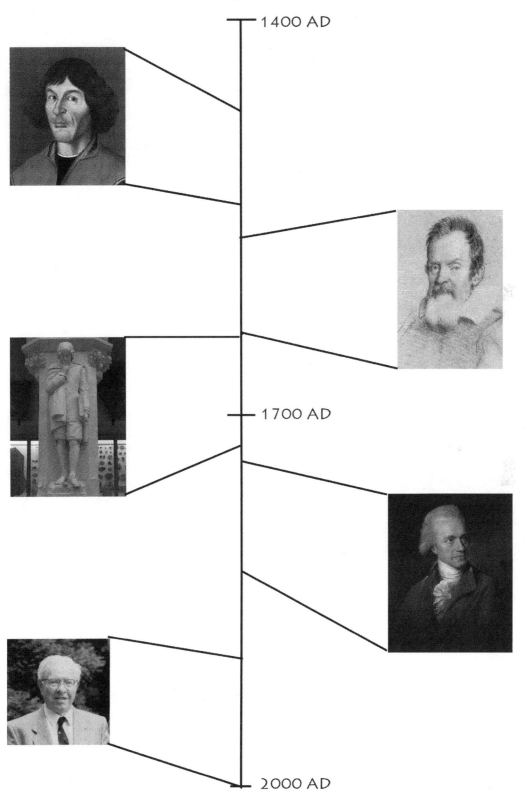

Student Guide Astronomy Unit 3: Astronomers & Their Tools ~ Week 12: Astronomers

Science Fair Project Step 1: Choose a Topic

Key 1 ~ Decide on an area of science.

What areas of earth science or astronomy are you interested in learning about?

_____
_____
_____
_____

Rank your interest in the different areas you listed and then circle the one area that you would like to use for your topic.

Key 2 ~ Develop several questions about the area of biology.

What questions would you like to answer about your area of earth science or astronomy?

(*Note* – Remember that good questions begin with how, what, when, who, which, why, or where.)

_____
_____
_____
_____
_____
_____

Key 3 ~ Choose a question to be the topic.

Write down the question that you will be using for your project.

_____
_____
_____

Science Fair Project Step 2: Do Some Research

Key 1 ~ Brainstorm for research categories.

What categories are you going to research for your project?

1. _____

2. _____

3. _____

4. _____

5. _____

Key 2 ~ Research the categories.

Use the following template for your research cards:

```
┌─────────────────────────────────────────────────┐
│ Category Number              Reference Letter   │
│                                                 │
│                                                 │
│            One piece of Information             │
│                                                 │
│                                                 │
│                                                 │
└─────────────────────────────────────────────────┘
```

Record your sources below.

A. _____

_____

B. _____

_____

C. _____

_____

D. _____

_____

E. _____

_____

F. _____

Written Assignment Week 12

Discussion Questions
1. Who first studied the stars?
2. What was early astronomy used for?
3. What was the Ptolemaic theory?
4. When did modern astronomy begin?
5. What is the difference between practical and theoretical astronomy?
6. Where can telescopes be placed?

Written Assignment Week 12

Student Assignment Sheet Astronomy Week 13
Looking Into Space

Experiment: Science Fair Project

This week, you will complete steps two through four of your Science Fair Project. You will be finishing your research, formulating your hypothesis, and designing your experiment.

2. **Do Some Research** – This week, you will finish your research. Then, organize your research index cards and write a brief report on what you have found out.
3. **Formulate a Hypothesis** – A hypothesis is an educated guess. For this step, you need to review your research and make an educated guess about the answer to your question. A hypothesis for the question asked in step one would be, "The more a comet weighs, the quicker it will melt."
4. **Design an Experiment** – Your experiment will test the answer to your question. You need to have a control and several test groups. Your control will have nothing changed, while your test groups will change only one factor at a time. An experiment to test the hypothesis given above would be to fill three different sizes of balloons—large, medium, and small—with varying amounts of water. Make sure you have three of each size to make the test valid and then freeze each one. Once they are frozen, apply a heat source for a set amount of time. Each time, measure how much water has melted and record.

Vocabulary & Memory Work
- Vocabulary: radio telescope, reflecting telescope, refracting telescope
- Memory Work – Continue to work on memorizing the Ten Nearest Galaxies and Their Types.

Sketch Assignment: Refracting & Reflecting Telescopes
- Label the following: Refracting Telescope, eye, object lens, focus, eyepiece lens and draw the light rays; Reflecting Telescope, eye, eyepiece lens, focus, object mirror, secondary mirror and draw the light rays

Writing Assignment
- Reading Assignment: *Kingfisher Science Encyclopedia* pg. 416-417 Looking Into Space
- Additional Research Readings:
    - The Optical Telescope: *DK Astro* pg. 24-25
    - The Radio Telescope: *DK Astro* pg. 28-29
    - Exploring Space: *KSE* pg. 418-419

Dates to Enter
- 1609 – Galileo invents the first telescope.
- 1931 – American astronomer, Karl Janley, collects the first evidence of radio radiation coming from space.
- 1992 – *Keck I*, the first telescope to use a segmented mirror, is completed.

Sketch Assignment Week 13

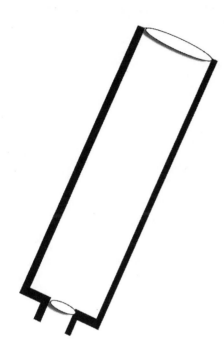

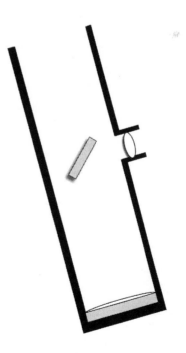

Student Guide Astronomy Unit 3: Astronomers & Their Tools ~ Week 13: Looking Into Space

Science Fair Project Step 2: Do Some Research

Key 3 ~ Organize the information.

Organize the information for your report.

Key 4 ~ Write a brief report.

Write down what the order of your categories will be for your report.

1. _____
2. _____
3. _____
4. _____
5. _____

On a separate sheet of paper write out a rough draft of your research report.

Science Fair Project Step 3: Formulate a Hypothesis

Key 1 ~ Review the Research.

Read over your research.

Key 2 ~ Formulate an Answer.

Write down your hypothesis for your science fair project.

_____
_____
_____
_____

Science Fair Project Step 4: Design an Experiment

Key 1 ~ Choose a Test.

What are some ways that you can test your hypothesis?

_____
_____
_____

_____
_____
_____

Key 2 ~ Determine the Variables.

What factor are we trying to test? (Independent variable)

_____

What factor will we use to measure the progress of our test? (Dependent variable)

_____

What factors do we need to keep constant so that they will not affect our results? (Controlled variables)

_____
_____

Key 3 ~ Plan the Experiment.

What will the groups in your experiment be?

Control Group: _____

Test Group 1: _____

Test Group 2: _____

Test Group 3: _____

Test Group 4: _____

Write down the plan for your experiment.

_____
_____
_____
_____
_____
_____

Student Guide Astronomy Unit 3: Astronomers & Their Tools ~ Week 13: Looking Into Space

Written Assignment Week 13

Discussion Questions

1. What do telescopes do for astronomers?
2. How do telescopes work?
3. What two modern advances do astronomers use to help them view space?
4. Why are telescopes placed on mountain tops?
5. What can telescopes in space see that we cannot spot on Earth?

Written Assignment Week 13

Student Assignment Sheet Astronomy Week 14
Exploring Space

Experiment: Science Fair Project

This week, you will complete steps five and six of your Science Fair Project. You will carry out the experiment and record your observations and results.

5. **Perform the Experiment** – This week, you will perform the experiment you designed last week. Be sure to take pictures along the way as well as record your observations and results. (**Note**—*Observations are a record of the things you see happening in your experiment. An observation would be "the small comet melted completely, but the large one remained mostly intact." Results are specific and measureable. Results would be that you collected 40 mL of water from the small comet. Observations are generally recorded in journal form, while results can be compiled into tables, charts, and graphs or relayed in paragraph form.*)

6. **Analyze the Data** – Once you have compiled your observations and results, you can use them to answer your question. You need to look for trends in your data and make conclusions from that. A possible conclusion to the electrolysis experiment would be, "Grass needs light to grow. The more light that grass is exposed to the better it will grow." If your hypothesis does not match your conclusion or your were not able to answer your question using the results from your experiment, you may need to go back and do some additional experimentation.

Vocabulary & Memory Work
- [ ] Vocabulary: space probe, rocket, space shuttle
- [ ] Memory Work – Continue to work on memorizing the Ten Nearest Galaxies and Their Types.

Sketch Assignment: Space Traveling Vehicles
- Label the following: *Saturn V, Titan III, Ariane V, Long March, Soyuz, H-IIA, US Space Shuttle*

Writing Assignment
- Reading Assignment: *Kingfisher Science Encyclopedia* pp. 420-421 Rockets and Space Planes
- Additional Research Readings
  - Venturing Into Space: *DK Astro* pp. 34-35
  - Humans in Space: *KSE* pp. 422-423

Dates to Enter
- 1926 – Robert Goddard launches the first liquid fuel rocket.
- 1942 – The first rocket, launched by Germany, reaches space.
- 1981 – NASA launches the first reusable space shuttle.
- 2011 – NASA's four reusable space shuttles are retired.

Sketch Assignment Week 14

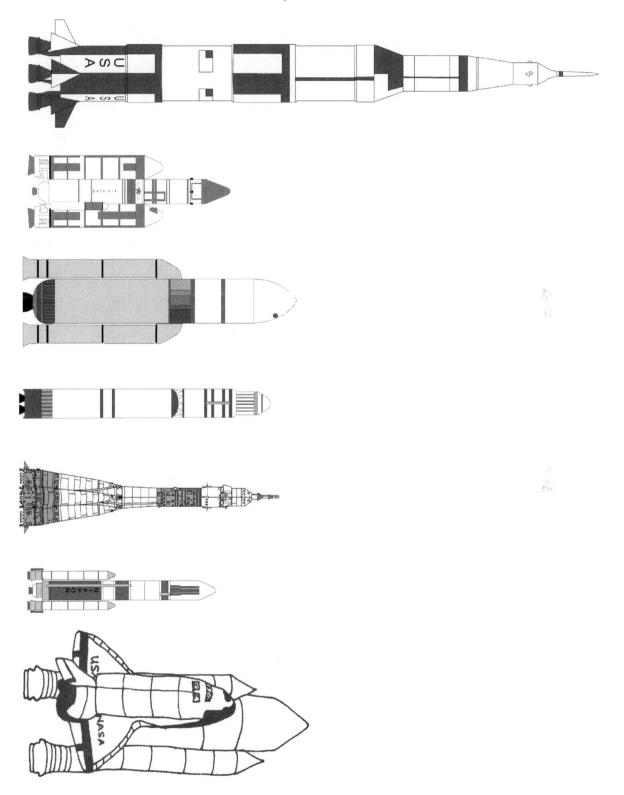

Student Guide Astronomy Unit 3: Astronomers & Their Tools ~ Week 14: Exploring Space

Science Fair Project Step 5: Perform the Experiment

Key 1 ~ Get ready for the experiment.

When do you plan to run your experiment?

From _____ to _____.

Purchase and gather your materials, and prepare any of the materials that need to be pre-made.

Key 2 ~ Run the experiment.

What things do you need to remember to do each day?

Take pictures of the experiment every day or for every trial.

_____
_____
_____
_____
_____
_____
_____
_____
_____

Key 3 ~ Record any Observations and Results.

Record your observations and results on a separate sheet of paper.

Science Fair Project Step 6: Analyze the Data

Key 1 ~ Review and organize the data.

What trends did you recognize in your observations?

_____
_____
_____

What information did you interpret from your results?

Key 2 ~ State the answer.

After reviewing your data, write the answer to your question. (**Note**—*Your statement should begin with "I found that…" or "I discovered that…"*)

Key 3 ~ Draw several conclusions.

Answer the following questions:
- ✓ Was my hypothesis proven true? (**Note**—*If your hypothesis was proven false, be sure to state why you think it was proven false.*)
- ✓ Did you have any problems or difficulties when performing your experiment?
- ✓ Did anything interesting happen that you would like to share?
- ✓ Can you think of any other things related to your project that you would like to test in the future?

Now take your answer from key two and your answers from key three to write your conclusion on a separate sheet of paper. Your paragraph should be four to six sentences in length.

Written Assignment Week 14

Discussion Questions

1. What is escape velocity?
2. What are two downsides of using a rocket?
3. What is the main difference between a rocket and the space shuttle?
4. What are the three main parts of the Space Shuttle?
5. Explain the 4 steps of a space shuttle mission.

Written Assignment Week 14

Student Assignment Sheet Astronomy Week 15
Satellites

Experiment: Science Fair Project

This week, you will complete steps seven and eight of your Science Fair Project. You will be writing and preparing a presentation of your Science Fair Project.

7. **Create a Board** — This week, you will be creating a visual representation of your science fair project that will serve as the centerpiece of your presentation. You will begin by planning the look of your board, then, move onto preparing the information, and finally, you will pull it all together.

8. **Give a Presentation** — After you have completed your presentation board, determine if you would like to include part of your experiment in your presentation. Then, prepare a 5 minute talk about your project. Be sure to include the question you tried to answer, your hypothesis, a brief explanation of your experiment, and the results plus the conclusion to your project. Be sure to arrive on time for your presentation. Set up your project board and any other additional materials. Give your talk and then ask if there are any questions. Answer the questions and end your time by thanking whoever has come to listen to your presentation.

Vocabulary & Memory Work
- Vocabulary: natural satellite, artificial satellite
- Memory Work – Continue to work on memorizing the Ten Nearest Galaxies and Their Types.

Sketch Assignment
- There is no sketch assignment this week.

Writing Assignment
- Reading Assignment: *Kingfisher Science Encyclopedia* pg. 424-425 Satellites
- Additional Research Readings
  - Space Exploration: *USE* pg. 174-177

Dates to Enter
- 1957 – A rocket delivers the first Russian satellite, *Sputnik*, into space.
- 1989 – The satellite *Hipparcos* is launched. Its job is to map the night sky.

Written Assignment Week 15

Discussion Questions
1. Once an artificial satellite is placed into orbit, what happens?
2. What are the four main types of orbits for artificial satellites?
3. How do artificial satellites get power?
4. What are the purposes of artificial satellites?
5. How do communication satellites work?

Science Fair Project Step 7: Create a Board

Key 1 ~ Plan out the board.

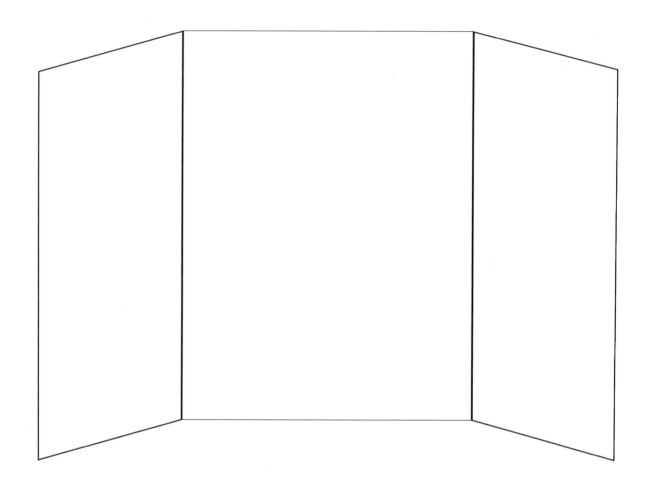

Key 2 ~ Prepare the information.

Type up the following information for your presentation board:

- ☐ Introduction
- ☐ Hypothesis
- ☐ Research
- ☐ Materials
- ☐ Procedure
- ☐ Results
- ☐ Conclusion

Key 3 ~ Put the board together.

☐ Put the decorative elements on your project board.

☐ Print out and attach your information paragraphs.

☐ Add the title to your science fair project board.

Science Fair Project Step 8: Give a Presentation

Key 1 ~ Prepare the presentation.

Write down the outline for your presentation on a separate sheet of paper.

Key 2 ~ Practice the presentation.

☐ Practice your presentation in front of a mirror several times.

☐ Practice your presentation with your teacher.

Key 3 ~ Share the presentation.

Keep the following tips in mind for your presentation:

- ✓ Arrive on time for your presentation.
- ✓ Set up your project board and any other additional materials.
- ✓ Give your talk and then ask if there are any questions.
- ✓ Answer the questions and end your time by thanking whoever has come to listen to your presentation.

Earth Science Unit 4
Our Planet

Earth Science Unit 4: Our Planet
Vocabulary Sheet

Define the following terms as they are assigned on your Student Assignment Sheet.

1. Mantle – _____

2. Cartographer – _____

3. Lines of Latitude – _____

4. Lines of Longitude – _____

5. Delta – _____

6. Estuary – _____

7. Source – _____

8. Coast – _____

9. Oceanic ridge – _____

10. Oceanic trench –

11. Tides –

12. Deposition –

13. Erratics –

14. Glacier –

15. Moraine –

16. Natural Cycle –

17. Greenhouse Gas –

18. Biome –

Student Assignment Sheet Week 16
Inside the Earth

Experiment: Model Earth
Materials
- ✓ Modeling clay (you will need yellow, orange, red, blue, and green)
- ✓ Ruler

Procedure
1. Read the introduction to this experiment.
2. Begin by making a ball about 1.2 cm across out of the yellow clay. This represents the Earth's inner core.
3. Then make another layer about 3 cm across out of the red clay around the ball. This layer represents the Earth's outer core.
4. Then you make another layer about 6 cm across out of the orange clay around the ball. This layer represents the Earth's mantle.
5. Finally, make some flattened pieces of blue and green clay to represent the Earth's crust and layer them over your ball.
6. Cut your ball in half and observe the layers of the Earth that were created.
7. Take a picture and complete your experiment sheet.

Vocabulary & Memory Work
- ☐ Vocabulary: mantle
- ☐ Memory Work – Work on memorizing the Major Lines of Longitude & Latitude.
    1. Prime Meridian
    2. Equator
    3. Tropic of Cancer
    4. Tropic of Capricorn

Sketch Assignment: Inside the Earth
- Label the Following: inner core, outer core, mantle, crust, ocean, and continent

Writing Assignment
- Reading Assignment: *The Kingfisher Science Encyclopedia* pp. 8-9 Earth's Structure
- Additional Research Readings
    - Earth's Structure: *USE* pp. 180-181
    - Structure of the Earth: *DKEOS* pp. 212-213

Dates
- 1965 – Tuzo Wilson, a Canadian geophysicist, explains how the plates on the ocean floor move.

Sketch Assignment Week 16

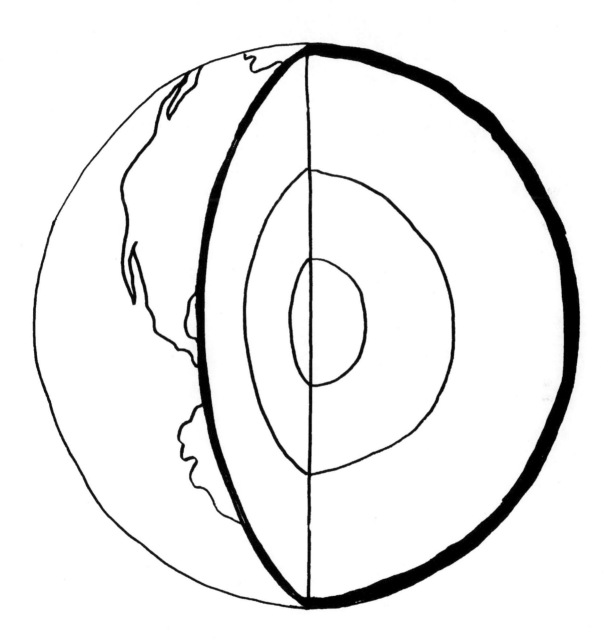

Student Guide Earth Science Unit 4: Our Planet ~ Week 16 Inside the Earth

Model Earth

Introduction

The Earth consists of three main layers, the crust, the mantle and the core. Some of these layers are solid, some of them are molten. Each layer varies in thickness and consists of different elements. In this experiment, you are going to use modeling clay to create a model of the Earth and its internal layers.

Materials

Procedure

Observations and Results

```
                          Picture of my Earth
┌─────────────────────────────────────────────────────────────┐
│                                                             │
│                                                             │
│                                                             │
│                                                             │
│                                                             │
│                                                             │
│                                                             │
│                                                             │
│                                                             │
└─────────────────────────────────────────────────────────────┘
```

Conclusion

_____
_____
_____
_____
_____
_____
_____

Written Assignment Week 16

Discussion Questions

1. What are the three main layers of the Earth? (*Be sure to include a detail or two about each.*)
2. How is the Earth's magnetic field created and why is it important?
3. What causes seismic activity on the Earth?
4. What have seismic scans of the Earth revealed about the mantle?

Written Assignment Week 16

Student Assignment Sheet Week 17
Maps and Mapping

Experiment: Is a flat representation the same as a spherical representation of the Earth?
  Materials
  ✓ Blue balloon (with the continents drawn or printed on it)
  ✓ Flat map
  ✓ Pin

  Procedure
  1. Read the introduction to this experiment and answer the question.
  2. Blow up your balloon so that it looks like a globe. (If your balloon did not have the continents already printed on it, now is the time to draw them.)
  3. Take your flat map and wrap it around the balloon, while attempting to match the continents up. Record your observations.
  4. Take the pin, pop your balloon and then cut it between The Americans and Asia. Try to stretch your balloon to match the map. Record your observations.
  5. Draw conclusions and complete your experiment sheet.

Vocabulary & Memory Work
  ☐ Vocabulary: cartographer, lines of longitude, lines of latitude
  ☐ Memory Work – Work on memorizing the Major Lines of Longitude & Latitude.
   1. Prime Meridian
   2. Equator
   3. Tropic of Cancer
   4. Tropic of Capricorn

Sketch: Lines on the Globe
  ▦ Label the Following: Lines of longitude, Lines of latitude, Arctic Circle, Antarctic Circle, Prime Meridian, Equator, Tropic of Cancer, Tropic of Capricorn, North Pole, South Pole

Writing
  ⌘ Reading Assignment: *The Kingfisher Science Encyclopedia* pp. 46-47 Maps and Mapping
  ⌘ Additional Research Readings
   📖 Mapping the Earth: *DKEOS* pg. 240

Dates
  🕒 1538 – Gerhard Mercator, a Flemish cartographer, devises a fairly accurate way to represent the Earth's surface on a map.

Sketch Assignment Week 17

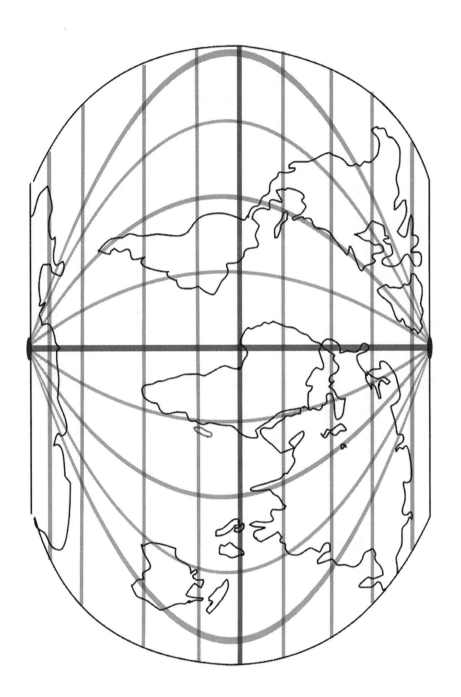

Student Guide Earth Science Unit 4: Our Planet ~ Week 17 Maps and Mapping

Is a flat representation the same as a spherical representation of the Earth?

Introduction

Maps are graphical representations of the Earth. There are several different types of maps, such as world maps, relief maps, topographical maps, and street maps, but they all have one thing in common. All maps share information that helps the user to determine position, so it is very important that they are accurate. The Earth is a spherical object, but most maps are flat representations. In this experiment, you are going to examine if flat representations of the Earth are the same as spherical representations.

Hypothesis

Is a flat representation the same as a spherical representation of the Earth?

               Yes   No

Materials

_____   _____

_____   _____

_____   _____

_____   _____

Procedure

_____

_____

_____

_____

_____

_____

_____

_____

Picture of my Earth

Observations and Results

Conclusion

Written Assignment Week 17

Discussion Questions
1. What is the purpose of a map?
2. What is the purpose of the lines that you find on a globe?
3. How are lines of latitude measured?
4. How do lines of longitude differ from lines of latitude?
5. How do cartographers represent a round Earth on a flat surface?
6. How are satellites used in mapping?

Written Assignment Week 17

Student Assignment Sheet Week 18
Rivers

Experiment: How is the course of a river determined?

Materials
- ✓ Pitcher for water
- ✓ Water
- ✓ Cookie sheet
- ✓ Paper cup
- ✓ Straw
- ✓ Dirt or sand
- ✓ Small rocks
- ✓ Books
- ✓ Tape

Procedure
1. Read the introduction to this experiment and answer the question.
2. Poke a hole in the base of the cup, just large enough for the straw to fit tightly, using a pencil. Insert the straw and place the cup in the sink. Fill the cup halfway with water and watch what happens. If the water comes out of the straw like a stream, the cup is ready to use. If not, make the necessary adjustments.
3. Using several books, prop your cookie sheet up at an angle, so that water will flow down easily. Tape the cup to the top of the cookie sheet and fill it three quarters full with water. Observe what happens and record where the water went.
4. Dry the cookie sheet off and then sprinkle some dirt or sand all over the sheet. Refill the cup three quarters full with water, observe what happens and record where the water went.
5. Clean the cookie sheet off, sprinkle some dirt or sand all over the sheet, and place a few rocks in various places. Refill the cup three quarters full with water, observe what happens and record where the water went.
6. Draw conclusions and complete your experiment sheet.

Vocabulary & Memory Work
- ☐ Vocabulary: delta, drainage basin, estuary, source
- ☐ Memory Work – Begin to work on the World's Major Seas & Oceans (*See the unit memory work on pg. 244 for a complete list.*)

Sketch Assignment: A River's Course
- Label the Following: source, upper stage, middle stage, tributary, lower stage, delta

Writing Assignment
- Reading Assignment: *The Usborne Science Encyclopedia* pp. 190-191 Rivers
- Additional Research Readings
  - Rivers: *DKEOS* pg. 233

Dates
- 3300 BC – The Indus Valley Civilization uses rivers for navigation.

Sketch Assignment Week 18

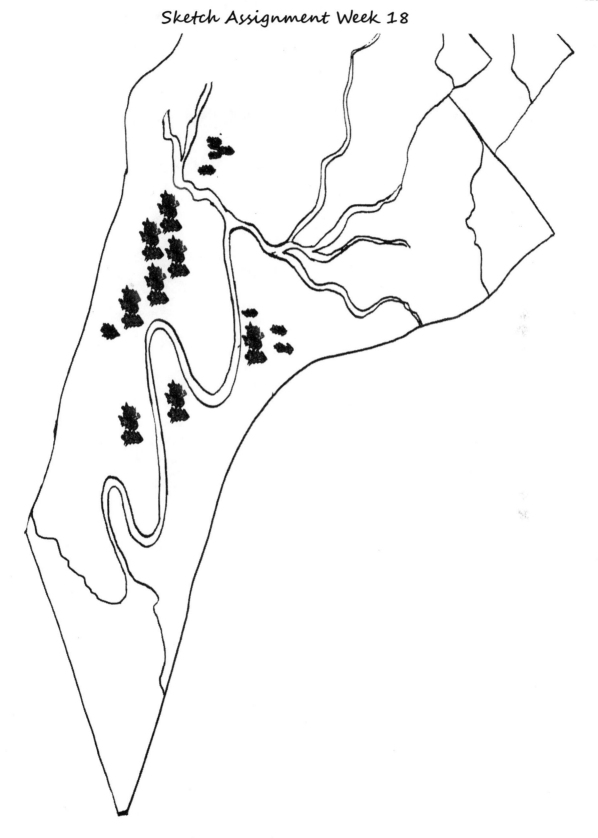

Student Guide Earth Science Unit 4: Our Planet ~ Week 18 Rivers

How is the course of a river determined?

Introduction

    A river is a natural flow of water from one area to a lake, sea, or ocean. Rivers typically contain fresh water and they are a very important part of the water cycle. Rivers are also home to a variety of wildlife and plant species. In this experiment you are going to examine how a rivers course is determined.

Hypothesis

I think that the course of a river is determind by _____

_____

_____

_____

Materials

Procedure

Observations and Results

Stream	Observations
Plain Stream	
Stream with Dirt	
Stream with Dirt & Rocks	

Conclusion

Written Assignment Week 18

Discussion Questions
1. Where does the water in a river come from?
2. Describe a river's course.
3. How do rivers erode the landscape around them?
4. How is a load deposited by a river?
5. Where is a river delta found and what happens at this point in the river?

Written Assignment Week 18

Student Assignment Sheet Week 19
Oceans

Experiment: Can surface currents affect the ocean floor the same way that deep water currents do?

Materials

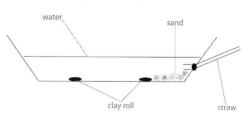

- ✓ Aluminum bread pan or Plastic bin
- ✓ Air dry clay
- ✓ Water
- ✓ Sand (1 cup)
- ✓ 2 Straws

Procedure

1. Read the introduction to this experiment and answer the question.
2. Have your teacher cut a small hole at the center of one end of your pan that is 1½ inches from the bottom. Insert your straw into the hole, so that when you blow through it the air skips over the bottom of the pan. Use some of the air dry clay to hold the straw in place and prevent water from coming out of the hole.
3. Next, roll out two lengths of air dry clay that are the exact width of the pan and place them 2 inches from the each end of the pan. (*See diagram above for a visual reference.*)
4. Fill your pan two-thirds of the way with water and add ½ cup of sand to the end of the pan that has the straw. Make sure that the sand stays between the air dry clay roll and the end of the pan.
5. Make some deep water currents by blowing air through the straw at the end of the pan intermittently over one minute. Observe what happens and record your results.
6. Empty the pan, clean it out, refill it one third of the way with water and add ½ cup of sand in the same way you did in step #4.
7. Make some surface currents by blowing air through the a straw onto the surface of the water in the pan intermittently over one minute. Observe what happens and record your results.
8. Draw conclusions and complete your experiment sheet.

Vocabulary & Memory Work
- ☐ Vocabulary: coast, oceanic ridge, oceanic trench, tides
- ☐ Memory Work – Continue to work on memorizing the World's Major Seas and Oceans.

Sketch Assignment: The World's Major Seas & Oceans
- 🖼 Label the Following: Arctic Ocean, Pacific Ocean, Atlantic Ocean, Indian Ocean, Southern Ocean, Mediterranean Sea, Red Sea, Black Sea, Caribbean Sea, Gulf of Mexico, Hudson Bay, Bering Sea, Tasman Sea, Coral Sea, Bay of Bengal, Arabian Sea, and North Sea

Writing Assignment
- ৬ Reading Assignment: *The Kingfisher Science Encyclopedia* pp. 12-13 The Oceans
- ৬ Additional Research Readings
    - 📖 Seas and Oceans: *USE* pp. 188-189, *DKEOS* pg. 234
    - 📖 The Shoreline: *DKEOS* pp. 236-237
    - 📖 The Ocean Floor: *KSE* pp. 14-15

Dates
- 🕒 1800 BC – Egyptians begin using very simple techniques to measure water depths.
- 🕒 1620 – Dutch physician, Cornelis Drebbel, builds the world's first submarine and makes several trips in the River Thames near London at a depth of about 12 or 15 feet.
- 🕒 1960 – A two manned submarine, named *Trieste*, dives to what was believed to be the deepest point in the Mariana Trench, which was 10,915 meters.

Sketch Assignment Week 19

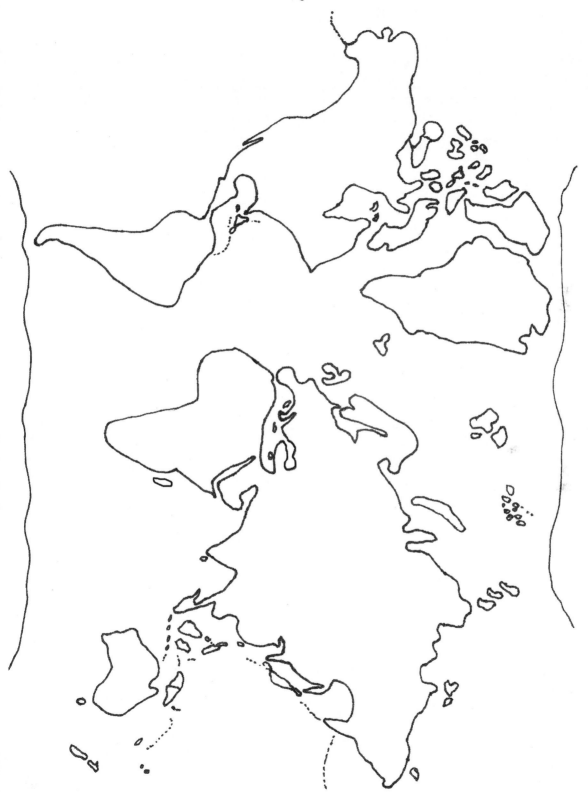

Student Guide Earth Science Unit 4: Our Planet ~ Week 19 Oceans

Can surface currents affect the ocean floor the same way that deep water currents do?

Introduction

The ocean is constantly in motion due to several currents that affect it. These currents are caused by several factors, such as wind, density, and temperature. Surface currents are caused by the wind blowing over the surface of the ocean, which then forms waves. Deep water currents are caused by cold water sinking and warm water rising. Deep water currents can also cause underwater waves. In this experiment, you are going to examine how these two different currents can affect the ocean floor.

Hypothesis

Can surface currents affect the ocean floor the same way that deep water current do?

        Yes   No

Materials

_____   _____
_____   _____
_____   _____
_____   _____

Procedure

_____
_____
_____
_____
_____
_____
_____
_____
_____

Observations and Results

Currents	Observations
Deep water currents	
Surface currents	

Conclusion

Written Assignment Week 19

Discussion Questions
1. How much of the Earth's surface is covered with water?
2. How do currents work in the ocean?
3. Why are the ocean depths so difficult to explore?
4. Where do waves form and why do they crash on the shore?

Written Assignment Week 19

Student Assignment Sheet Week 20
Glaciers

Experiment: What happens when glaciers melt?
  Materials
  - Glacier Melt Model (*See note below for how to make using cup, water, pebbles, and sand.*)
  - Large cutting board with a handle
  - Large rubber band

  Procedure

  NOTE – *The day before you do this experiment, you will need to make your Glacier Melt Model. Begin by pouring a layer of both pebbles and sand on the bottom of your cup, then cover with water and place in the freezer. Once the water has frozen, repeat the process again until the cup has been filled. You should have at least three layers.*

  1. Read the introduction to this experiment and answer the question.
  2. Prop your cutting board up in the sink, handle-side up, so that the board is at about a 45 degree angle.
  3. Remove your Glacier Melt Model from the cup using a little warm water. Use the rubber band to secure it pebble/sand side down to the top of the cutting board. Observe what happens as the ice melts.
  4. Record your observations and the time it takes for the model to melt.
  5. Draw conclusions and complete your experiment sheet.

Vocabulary & Memory Work
- [ ] Vocabulary: deposition, erratics, glacier, moraine
- [ ] Memory Work – Continue to work on memorizing the World's Major Seas & Oceans.

Sketch Assignment: Anatomy of a Glacier
- Label the Following: Cirque, Crevasses, Glacier's Snout, Meltwater, Movement of the Glacier (including arrows)

Writing Assignment
- Reading Assignment: *The Kingfisher Science Encyclopedia* pp. 34-35 Glaciers and Ice Sheets
- Additional Research Readings
  - Ice and Glaciers: *DKEOS* pp. 34-35

Dates
- No dates to be entered this week.

Sketch Assignment Week 20

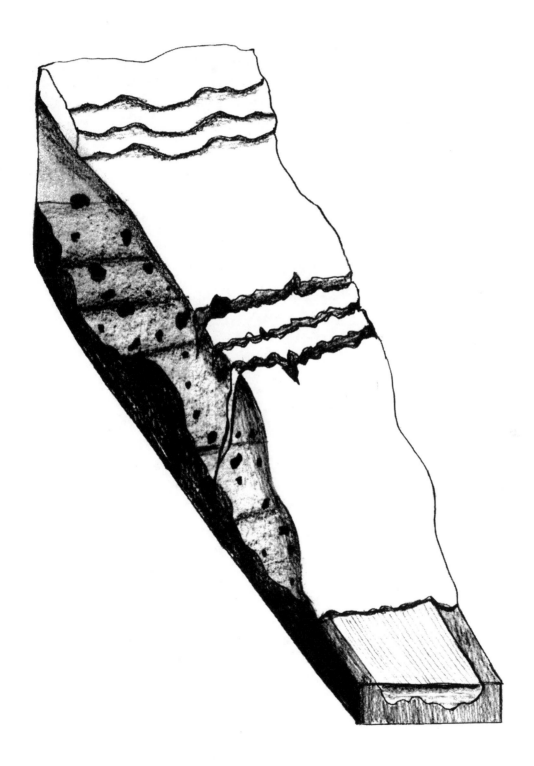

Student Guide Earth Science Unit 4: Our Planet ~ Week 20 Glaciers

What happens when glaciers melt?

Introduction

Glaciers are huge masses of ice that flow downhill. Glaciers are solid, powerful forces that can cut through the landscape, picking up rocks and soil along the way. In this experiment, you are going to look at what happens when those glaciers begin to melt.

Hypothesis

I think that when glaciers melt _____

_____

_____

_____

Materials

Procedure

Observations and Results

_____
_____
_____
_____
_____
_____
_____

Conclusion

_____
_____
_____
_____
_____
_____

Written Assignment Week 20

Discussion Questions
1. Where are glaciers found?
2. How do glaciers form?
3. What are some clues that let you know a glacier was once present in a valley?

Written Assignment Week 20

Student Assignment Sheet Week 21
Natural Cycles

Experiment: Can I recreate the water cycle?
   Materials:
   - ✓ Cup
   - ✓ Water
   - ✓ Plastic Baggie
   - ✓ Rubber Band

   Procedure:
   1. Read the introduction to this experiment and answer the question.
   2. Pour a little warm water into the bottom of your cup. Cover it with the plastic baggie and use the rubber band to secure it in place.
   3. Set the cup in a place where it will receive direct sunlight for the next two hours. (***NOTE** – If you do not have a sunny place to set your cup, set it under a desk lamp instead.*)
   4. Check the glass after 2 hours; observe and record what has happened.
   5. Draw conclusions and complete your experiment sheet.

Vocabulary & Memory Work
- ☐ Vocabulary: natural cycle, greenhouse gas
- ☐ Memory Work – Continue to work on memorizing the World's Major Seas & Oceans.

Sketch Assignment: The Nitrogen Cycle & The Carbon Cycle
- The Nitrogen Cycle – Label the following: Bacteria convert ammonia in the soil into nitrate, which is taken up by plants; plants also take up nitrogen from the air; plants are eaten by animals; as dead plants and animals decay, nitrogen is released into the soil.
- The Carbon Cycle – Label the following: When fossil fuels are burned, carbon dioxide is released into the air; plants take in carbon dioxide from the air to make food; animals take in carbon when they eat plants and release carbon dioxide when they breathe; as dead plants and animal decay, they release carbon dioxide into the air.

Writing Assignment
- Reading Assignment: *The Usborne Science Encyclopedia* pp. 292-293 Natural Cycles
- Additional Research Readings
  - Nutrient Cycles: *USE* pp. 334-335

Dates
- 1896 – Svante Arrhenius, a Swedish chemist, shows that CO_2 helps to trap heat in the atmosphere.

Sketch Assignment Week 21

Student Guide Earth Science Unit 4: Our Planet ~ Week 21 Natural Cycles

Can I Recreate the Water Cycle?

Introduction

The water cycle shows how water changes forms on the Earth.

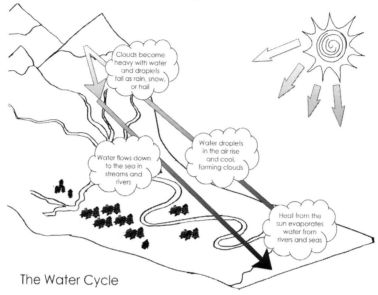

The Water Cycle

In today's experiment, you are going to try to create your very own mini-version of the water cycle.

Hypothesis

Can I recreate the water cycle?   Yes   No

Materials

_____   _____
_____   _____
_____   _____
_____   _____

Procedure

Observations and Results

[Picture of the mini-water cycle]

Conclusion

Written Assignment Week 21

Discussion Questions
1. Briefly describe the nitrogen cycle.
2. Briefly describe the carbon cycle.
3. Briefly describe the water cycle.
4. How are the natural cycles being upset?

Written Assignment Week 21

Student Assignment Sheet Week 22
Biomes and Habitats

Experiment: How do desert plants live with so little water?

Materials
- ✓ Paper towels
- ✓ Cookie sheet
- ✓ Wax paper
- ✓ Rubber band
- ✓ Water

Procedure
1. Read the introduction to this experiment and answer the question.
2. Moisten three paper towels with water until they are soaked, but not dripping, with water.
3. Lay one of the paper towels flat on the cookie sheet.
4. Roll the second paper towel up, secure it with the paper clips and lay it next to the flat paper towel on the cookie sheet.
5. Place the third paper towel flat on a piece of wax paper that is slightly larger, roll the two up together and secure with paper clips. Then, lay the roll next to the other papers towels on the cookie sheet and set the sheet in a place where it will not be disturbed.
6. Check your paper towels the next day, make observations, and record the results.
7. Draw conclusions and complete your experiment sheet.

Vocabulary & Memory Work
- ☐ Vocabulary: biome
- ☐ Memory Work – Continue to work on memorizing the World's Major Seas & Oceans.

Sketch Assignment: Biomes around the World
- Choose one of the following biomes—arctic, desert, grasslands, rainforest, taiga, or tundra. Find where the particular biome is on the globe and label that region on the globe.

Writing Assignment
- Reading Assignment: *The Kingfisher Science Encyclopedia* pp. 68-69 Biomes and Habitats
- Additional Research Readings
  - Plants and People: *USE* pp. 290-291
  - Major Biomes: *DKEOS* pp. 382-396

Dates
- 20th century – Tropical and temperate rainforests experience heavy logging, which leads to a reduction of the forests' sizes.

Sketch Assignment Week 22

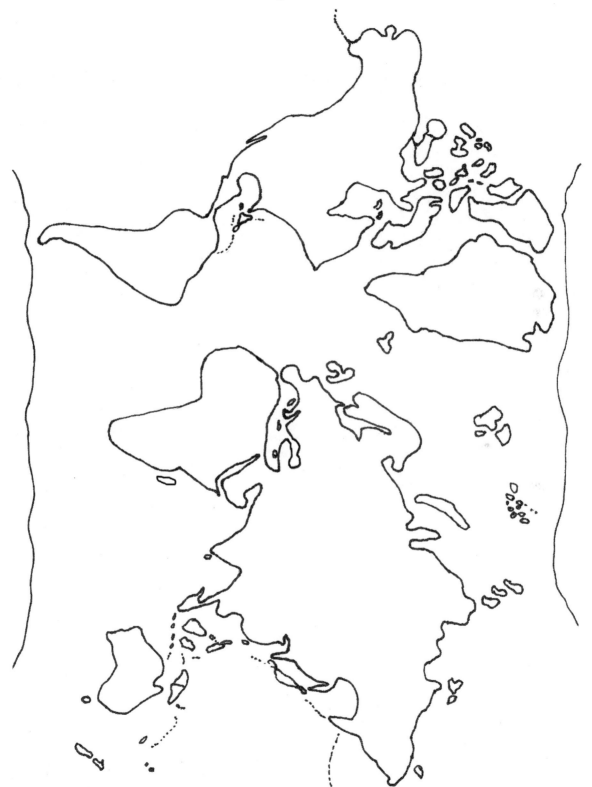

Student Guide Earth Science Unit 4: Our Planet ~ Week 22 Biomes and Habitats

How do desert plants live with so little water?

Introduction

The desert is a harsh environment in which only the hardiest plants and animals can survive. It is extremely hot during the day and very cold at night. The desert also receives very little rainfall. The plants and animals living there are well adapted to these conditions, like the camel, which can go for months without drinking water because its digestive system is extremely effective at extracting water. In this experiment, you are going to test how desert plants have adapted to survive with so little water.

Hypothesis

How do desert plants live with so little water?

_____

_____

_____

Materials

_____   _____

_____   _____

_____   _____

_____   _____

Procedure

_____

_____

_____

_____

_____

_____

_____

Student Guide Earth Science Unit 4: Our Planet ~ Week 22 Biomes and Habitats

Observations and Results

Sample	Observations the next day
Flat Paper Towel	
Rolled Paper Towel	
Rolled Paper Towel with Wax Paper	

Conclusion

Written Assignment Week 22

Discussion Questions
1. What determines the boundaries between biomes?
2. What is the difference between a habitat and a biome?
3. What is a community?

Written Assignment Week 22

Earth Science Unit 5
Geology

Earth Science Unit 5: Geology
Vocabulary Sheet

Define the following terms as they are assigned on your Student Assignment Sheet.

1. Continent – _____
   _____
   _____

2. Faults – _____
   _____
   _____

3. Magma – _____
   _____
   _____

4. Lava – _____
   _____
   _____

5. Spreading Ridge – _____
   _____
   _____

6. Subduction Zone – _____
   _____
   _____

7. Seismic Wave – _____
   _____
   _____

8. Richter Scale – _____
   _____
   _____

9. Mercalli Scale – _____
_____
_____

10. Mountain Range – _____
_____
_____

11. Fossil – _____
_____
_____

12. Strata – _____
_____
_____

13. Gem – _____
_____
_____

14. Ore – _____
_____
_____

15. Erosion – _____
_____
_____

16. Weathering – _____
_____
_____

Student Assignment Sheet Week 23
Continents

Experiment: What type of plate movement is responsible for making mountains?

Materials
- ✓ Marshmallow creme (or whipping cream)
- ✓ Graham crackers
- ✓ 3 Plates
- ✓ Bowl with about an inch of water

Procedure
1. Read the introduction to this experiment and fill in the hypothesis blank.
2. On each of the plates, add about half a cup of marshmallow creme and label the plates with #1, #2, and #3.
3. On the plate #1, you will recreate a divergent plate movement. Begin by heating the plate up for 10 seconds so that the fluff is a bit warm. Then, break one of the graham crackers in half and quickly dip both of the cracked edges into the bowl of water. Place both crackers on the marshmallow creme with the dipped ends next to each other. Gently push the two ends away from each other and observe what happens.
4. On the plate #2, you will recreate a convergent plate movement. Begin by heating the plate up for 10 seconds so that the fluff is a bit warm. Then, break one of the graham crackers in half and quickly dip both cracked edges into the bowl of water. Place both crackers on the marshmallow creme with the dipped ends next to each other. Gently push the two ends towards each other and observe what happens.
5. On the plate #3, you will recreate a transforming plate movement. Begin by heating the plate up for 10 seconds so that the fluff is a bit warm. Then, break one of the graham crackers in half and quickly dip both cracked edges into the bowl of water. Place both crackers on the marshmallow creme with the dipped ends next to each other. Gently push one cracker up and the other cracker down, so that the ends slide past each other. Observe what happens.
6. Write down your observations, draw conclusions, and complete your experiment sheet.

Vocabulary & Memory Work
- ☐ Vocabulary: continents, faults
- ☐ Memory Work – Work on memorizing the Tectonic Plates. (*See pg. 245 for a complete listing.*)

Sketch: The Seven Continents
- Label the Following: North America, South America, Europe, Asia, Africa, Australia (or Oceania), Antarctica

Writing
- Reading Assignment: *The Kingfisher Science Encyclopedia* pp. 16-17 Continental Drift
- Additional Research Readings
  - Moving Continents: *DKEOS* pp. 214-215

Dates
- c. 30 AD – Prominent geographer, Strabo, suggests that there might be continents that are not yet known to the Greeks.

Sketch Assignment Week 23

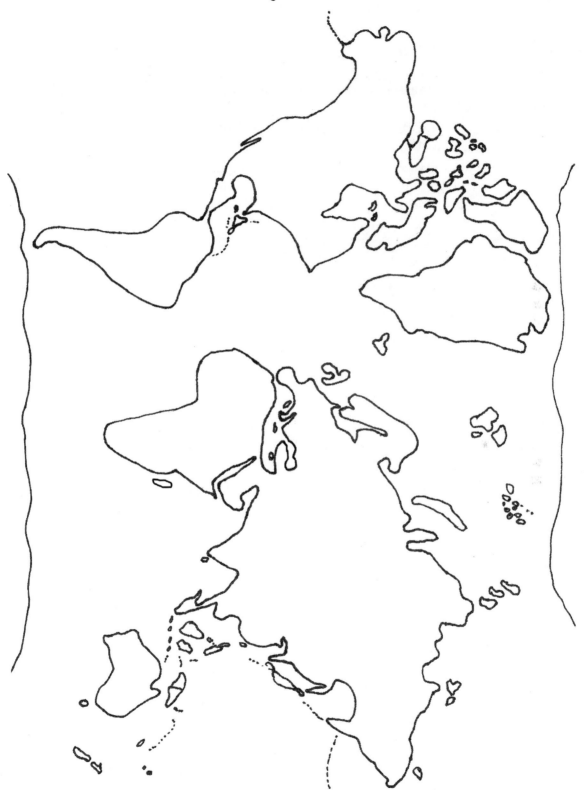

Student Guide Earth Science Unit 5: Geology ~ Week 23 Continents

What type of plate movement is responsible for making mountains?

Introduction

There are three main types of plate movements—divergent, convergent, and transforming. Each of these types of plate movements explain how the pieces of the Earth's crust interact with each other at the points at which they meet. In divergent plate movements, the two plates move away from each other. In convergent plate movements, the two plates move towards each other. In transforming plate movements, the two plates slide past each other. In this experiment, you are going to look at models of the three plate movements and see the results of their interactions.

Hypothesis

I think that _____ plate movement is responsible for making mountains.

      divergent   convergent   transforming

Materials

Procedure

Observations and Results

Plate Number	My Observations

Conclusion

Written Assignment Week 23

Discussion Questions

1. What creates the Earth's solid surface?
2. What causes the movement of the tectonic plates of the Earth's crust?
3. What is one difference between oceanic and continental crust?
4. What happens as the tectonic plates stretch? Collide?
5. How can geologists track the movement of the tectonic plates?

Written Assignment Week 23

Student Assignment Sheet Week 24
Volcanoes

Experiment: Exploding Volcano

Materials
- ✓ Mentos™
- ✓ Cardboard cereal box
- ✓ 1-Liter bottle of cola (or orange soda)
- ✓ 1 Can of Great Stuff™ Foam
- ✓ Paints
- ✓ Aluminum Foil

> ☢ **CAUTION**
> Please follow all the safety instructions on the bottle of Great Stuff™ when forming the volcano.

Procedure
1. Read the introduction to this experiment.
2. Cover your bottle with aluminum foil so that it fits snugly over the bottle. Then, have an adult spray Great Stuff™ around the bottle, starting at the bottom to make a cone volcano.
3. Let the foam dry for the recommended amount of time, remove the bottle (leaving the foil behind), and paint the foam to look like a volcano. Let the paint dry.
4. Meanwhile, cut a rectangle and a square out of your cereal box. Roll one of the rectangles into a tube that will fit three Mentos™ and tape it.
5. Take your supplies outside, remove the top from your soda bottle, and place the foam volcano over it. Cover the opening with the flat cardboard square, place your cardboard tube on top of that, and load three Mentos™ into it.
6. Quickly remove the flat cardboard square so the Mentos™ drop in the soda, step back, and observe what happens.
7. Take a picture and complete your experiment sheet.

Vocabulary & Memory Work
- ☐ Vocabulary: magma, lava, spreading ridge, subduction zone
- ☐ Memory Work – Continue to work on memorizing the Tectonic Plates.

Sketch Assignment: Cross-section of a Cone Volcano
- Label the Following: dust, ash, and gases, crater, lava flow, main vent, side vents, magma chamber

Writing Assignment
- Reading Assignment: *The Kingfisher Science Encyclopedia* pp. 18-19 Volcanoes
- Additional Research Readings:
  - 📖 Volcanoes: *DKEOS* pp. 216-217
  - 📖 Volcano Section: *USE* pp. 182-183

Dates
- 🕐 79 AD – Mount Vesuvius, in Italy, erupts and destroys Pompeii.
- 🕐 May 18, 1980 – Mount St. Helens, in Washington, erupts.

Sketch Assignment Week 24

Student Guide Earth Science Unit 5: Geology ~ Week 24 Volcanoes

A Cone Volcano

Introduction

Volcanoes are openings in the Earth's surface that are formed by magma bursting out. When they erupt, lava and ash are thrown out and build up along the sides. Eventually, the lava and ash set up and form a solid layer of volcanic rock. In this experiment, you are going to create your own volcanic eruption.

Materials

Procedure

Observations and Results

Picture of my Volcano

Conclusion

_____
_____
_____
_____
_____
_____
_____

Written Assignment Week 24

Discussion Questions
1. What is a volcano?
2. What does the behavior of a volcano depend on?
3. Name three different types of volcanos and share a characteristic about each.
4. What happens during a volcanic eruption?

Written Assignment Week 24

Student Assignment Sheet Week 25
Earthquakes

Experiment: Do seismic p-waves travel in the same way as seismic s-waves?

Materials
- ✓ Partner
- ✓ Slinky
- ✓ Rope

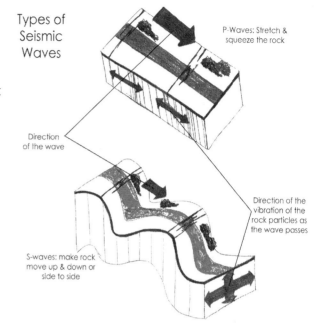

Types of Seismic Waves

Procedure
1. Read the introduction to this experiment and answer the question.
2. Sit down at a table across from your partner, set the slinky on the table, and stretch it gently between the two of you. Gently push your end of the slinky towards your partner several times and observe what happens. Record how the slinky moved on your experiment sheet and set the slinky aside.
3. Set the rope on the table and stretch it gently between the two of you. Move your end of the rope side to side several times and observe what happens. Record how the rope moved on your experiment sheet.
4. Draw conclusions and complete your experiment sheet.

Vocabulary & Memory Work
- ☐ Vocabulary: seismic wave, Richter Scale, Mercalli Scale
- ☐ Memory Work – Continue to work on memorizing the Tectonic Plates.

Sketch Assignment: Tectonic Plates Around the Globe
Label the Following: Juan de Fuco, North American, Cocos, Caribbean, South American, Nazca, Scotia, Antarctic, African, Arabian, Eurasian, Indo-Australian, Philippine, Carolina, Bismarc, Fiji, Pacific

Writing Assignment
- Reading Assignment: *The Kingfisher Science Encyclopedia* pp. 20-21 Earthquakes
- Additional Research Readings
  - Earthquakes: *USE* pg. 220

Dates
- 132 – The Chinese invent the first seismograph, in which precisely balanced metal balls fall if the ground shakes.
- 1935 – Charles Richter, a US seismologist, develops a scale for reporting the strength of earthquakes.

Sketch Assignment Week 25

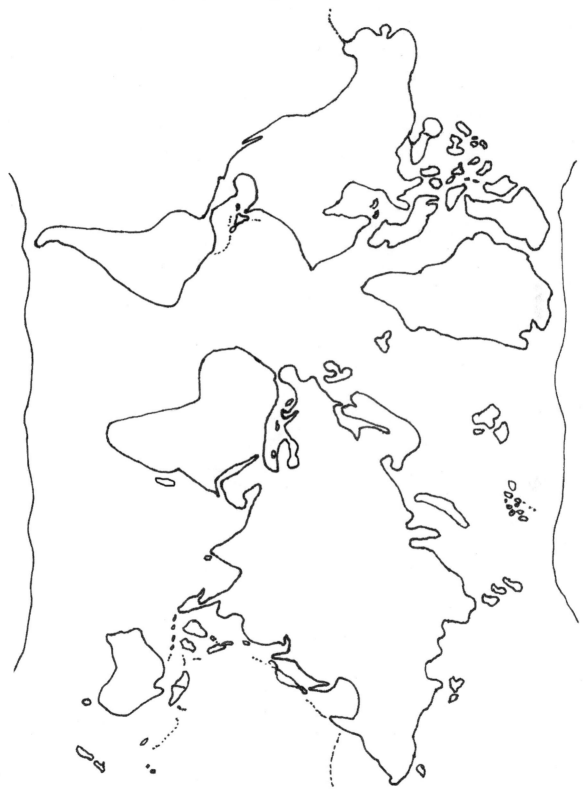

Student Guide Earth Science Unit 5: Geology ~ Week 25 Earthquakes

Do seismic p-waves travel in the same way as seismic s-waves?

Introduction

    Earthquakes happen when plates slide along each other and cause the rock to be twisted, stretched or squeezed. The movement of the rock is caused by seismic waves that are sent out from the focus of the earthquake. In this experiment, you are going to look at two types of seismic waves, p-waves (using a slinky) and s-waves (using a rope). You will also examine the differences and similarities of the two waves.

Hypothesis

Do seismic p-waves travel the same as seismic s-waves?

        Yes   No

Materials

Procedure

Observations and Results

	Description of the Movement
Slinky	
Rope	

Conclusion

Written Assignment Week 25

Discussion Questions
1. What is an earthquake?
2. Where do earthquakes most commonly occur?
3. What causes the destruction in a earthquake?
4. What are the two scales used to measure earthquakes?
5. Why is it so difficult to create a warning system for earthquakes?
6. What is a seismograph?

Written Assignment Week 25

Student Assignment Sheet Week 26
Mountains

Experiment: How does the rock cycle work?

Materials
- ✓ Several different colors of crayons
- ✓ Grater
- ✓ Butter knife
- ✓ Pencil sharpener
- ✓ Foil
- ✓ Bowl
- ✓ Warm water
- ✓ Foil muffin cups

Procedure
1. Read the introduction to this experiment.
2. Begin by using the grater, butter knife, and pencil sharpener to break apart, or "weather," a crayon. Catch the shavings, or "sediments," on a piece of aluminum foil. Once you have enough, gather up the "sediments" and fold the foil to create a pocket around them. Then, use a hand or a foot to compress the sediment shavings. Open up the foil packet and observe the changes.
3. Next, wrap the rock you created in the previous step in foil and place the packet in a bowl filled with warm water for one to two minutes or until the contents are soft. Take the foil packet out of the water and use a hand to massage and squeeze the rock into a ball. Open up the foil packet and observe the changes.
4. Finally, place the rock you created in the previous step in a foil muffin cup. Put the muffin cup in a tin and place it in an oven set to 300°F for 15 minutes. Take the tin out and allow the rocks to cool. Remove the rock from the foil cup and observe the changes.
5. Write down your observations, draw conclusions, and complete your experiment sheet.

Vocabulary & Memory Work
- ☐ Vocabulary: mountain range
- ☐ Memory Work – Begin to work on memorizing the Types of Rock. (*See pg. 244 for a complete list.*)

Sketch Assignment: Mountain Features
- Label the Following: nappe, recumbent fold, anticline, syncline, valley, fault, horst, rift valley

Writing Assignment
- Reading Assignment: *The Kingfisher Science Encyclopedia* pp. 22-23 Building Mountains
- Additional Research Readings
  - Mountain Building: *DKEOS* pp. 218-219

Dates
- 1785 – James Hutton, a Scottish geologist, says that moutains are formed by hot rock erupted from volcanoes.

Sketch Assignment Week 26

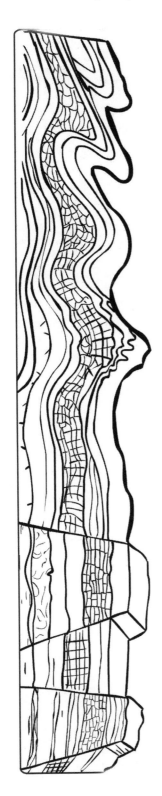

Student Guide Earth Science Unit 5: Geology ~ Week 26 Mountains

How does the rock cycle work?

Introduction

The rock cycle is a process that explains the changes rocks go through. It is summed up in the graphic below:

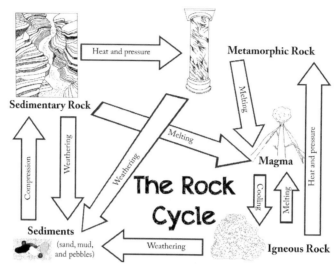

In today's experiment, you are going use crayons to see how the rock cycle works.

Materials

_____   _____
_____   _____
_____   _____
_____   _____

Procedure

_____
_____
_____
_____
_____
_____

Observations and Results

	My Observations
After Completing Step 2	
After Completing Step 3	
After Completing Step 4	

Conclusion

Written Assignment Week 26

Discussion Questions
1. What four factors contribute to the formation and destruction of mountains?
2. What are some of the newer mountain chains around the globe? Older ones?
3. What can the Earth's crust tell us about mountains?
4. Explain how the Himalayan Mountains were formed.

Written Assignment Week 26

Student Assignment Sheet Week 27
Rocks

Experiment: Rock Collection (week 1)

Materials
- ✓ 10 to 12 rocks collected from outside
- ✓ Rock & Mineral field guide
- ✓ Plastic baggie
- ✓ Sharpie Marker
- ✓ White-out

Procedure
1. Go outside and collect seven to ten different rocks.
2. Once inside, paint a small section of each rock, assign each one a number, and write that number on the painted section of the rock.
3. Then, use your field guide to identify the rocks you have collected. Write the identification of each rock, along with a fact or two about the rock, on the experiment sheet.
4. Place your rock samples in a plastic baggie. You will finish this project next week.

Vocabulary & Memory Work
- ☐ Vocabulary: fossil, strata
- ☐ Memory Work – Continue to work on memorizing the Types of Rock.

Sketch Assignment: Types of Rock
Label the following: igneous rock, metamorphic rock, sedimentary rock, mineral, fossil

Writing Assignment
- Reading Assignment: *The Kingfisher Science Encyclopedia* pg. 28 Igneous Rock, pg. 29 Metamorphic Rock, and pp. 30-31 Sedimentary Rock
- Additional Research Readings
  - Igneous Rock: *DKEOS* pg. 222
  - Sedimentary Rock: *DKEOS* pg. 223
  - Metamorphic Rock: *DKEOS* pg. 224
  - Fossils: *DKEOS* pg. 225

Dates
- 1546 – Georgius Agricola, a German metallurgist, first uses the term "fossil" when referring to the rocklike remains of animals and plants.

Sketch Assignment Week 27

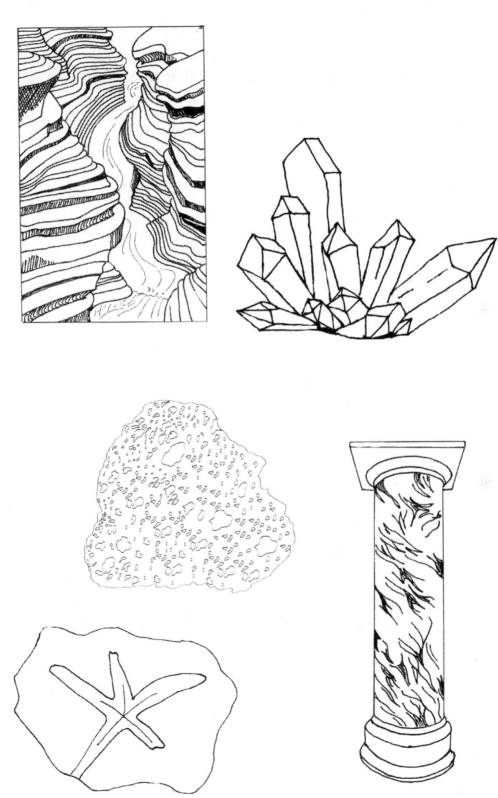

Student Guide Earth Science Unit 5: Geology ~ Week 27 Rocks

My Rock Collection (Week 1)

1. _____

   Interesting Facts: _____

   _____

   _____

2. _____

   Interesting Facts: _____

   _____

   _____

3. _____

   Interesting Facts: _____

   _____

   _____

4. _____

   Interesting Facts: _____

   _____

   _____

5. _____

   Interesting Facts: _____

   _____

   _____

6. _____

   Interesting Facts: _____

   _____

   _____

7. _____

   Interesting Facts: _____

   _____

   _____

8. _____

   Interesting Facts: _____

   _____

   _____

9. _____

   Interesting Facts: _____

   _____

   _____

10. _____

    Interesting Facts: _____

    _____

    _____

11. _____

    Interesting Facts: _____

    _____

    _____

12. _____

    Interesting Facts: _____

    _____

    _____

Written Assignment Week 27

Discussion Questions

Pg. 28 Igneous Rock
1. Describe igneous rock and how it is classified.
2. What are the two ways that igneous rock can form?

Pg. 29 Metamorphic Rock
1. What is metamorphic rock and how does it form?
2. Describe the two types of metamorphic rock.

Pp. 30-31 Sedimentary Rock
1. What is sedimentary rock?
2. What are the three types of sedimentary rock?
3. What are the two main ways of weathering rock (explain both)?

Written Assignment Week 27

Student Assignment Sheet Week 28
Ores and Gems

Experiment: Rock Collection (week 2)

Materials
- ✓ 10 to 12 rocks more collected from outside
- ✓ Rock & Mineral field guide
- ✓ Foam board
- ✓ Sharpie Marker
- ✓ White-out

Procedure
1. Go outside and collect five to eight more rocks.
2. Once inside, paint a small section of each rock, assign each one a number, and write that number on the painted section of the rock, just like you did for the previous week.
3. Then, use your field guide to identify the rocks you have collected. Write the identification of each rock, along with a fact or two about the rock on the experiment sheet.
4. Mount each rock, in order, on your foam board. You should have a total of 15 samples on your board. Once you have mounted the rocks, create a key somewhere on your board that identifies each type of rock.

Vocabulary & Memory Work
- ☐ Vocabulary: gem, ore
- ☐ Memory Work – Continue to work on memorizing the Types of Rocks.

Sketch Assignment: Rock Research Project

This week for the sketch, you will choose a rock, research that rock, and create a profile page for the rock. You can do this by completing the following steps:

1. ***Choose a rock*** – Choose a rock that you want to learn more about.
2. ***Do some research about the rock*** – Use the internet and the resources you have in your home or at your library to find out more about your chosen rock. As you do your research, write down any interesting facts and answer the following questions:
    - ✓ What group is your rock from? What is its composition or chemical formula? What is the typical crystal system for your rock? What is the typical color of your rock? What is the hardness rating for your rock? Does your rock cleave?
3. ***Complete the profile page for your rock*** – Write the name for your rock on the blank at the top of the page. Then fill out the remaining information. Your summary of interesting facts should include five to seven facts about the rock. Finally, draw your rock or glue a picture of it in the box.

Writing Assignment
- ✎ Reading Assignment: *The Kingfisher Science Encyclopedia* pp. 26-27 Ores and Gems
- ✎ Additional Research Readings
    - 📖 Rocks and Minerals: *DKEOS* pg. 221

Dates
- 🕐 1822 – Friedrich Mohs, a German mineralogist, creates the first hardness scale based on 10 common minerals to use as a reference.

Sketch Assignment Week 28 – Rock Profile Page

Group _____

Composition (Chemical Formula) _____

Crystal System _____

Color _____

Hardness _____

Cleavage _____

Interesting Facts

_____
_____
_____
_____
_____
_____
_____
_____
_____
_____
_____
_____

My Rock Collection (Week 2)

13. _____

   Interesting Facts: _____

   _____

   _____

14. _____

   Interesting Facts: _____

   _____

   _____

15. _____

   Interesting Facts: _____

   _____

   _____

16. _____

   Interesting Facts: _____

   _____

   _____

17. _____

   Interesting Facts: _____

   _____

   _____

18. _____

   Interesting Facts: _____

   _____

   _____

19. _____

   Interesting Facts: _____

   _____

   _____

20. _____

   Interesting Facts: _____

   _____

   _____

21. _____

   Interesting Facts: _____

   _____

   _____

22. _____

   Interesting Facts: _____

   _____

   _____

23. _____

   Interesting Facts: _____

   _____

   _____

24. _____

   Interesting Facts: _____

   _____

   _____

Written Assignment Week 28

Discussion Questions
1. Describe three ways that mineral deposits can be formed.
2. What is the purpose of smelting?
3. What are types of minerals make up gemstones?

Written Assignment Week 28

Student Assignment Sheet Week 29
Erosion and Weathering

Experiment: Do plants help to prevent erosion?
  Materials
  - ✓ Dirt or sand
  - ✓ Grass seed
  - ✓ Water
  - ✓ Pitcher
  - ✓ 2 Aluminum Pans

> **NOTE** – You will need to plant your grass seed mountain (steps two and three) at least 7 days before doing this experiment so that the grass will have the time to grow.

  Procedure
  1. Read the introduction to this experiment and answer the question.
  2. Build up a mountain of dirt in each of your pans. Be sure that both of your mountains are of equal height. Record the height on your experiment sheet.
  3. Set one mountain aside. Plant the other mountain with grass seed and water it frequently until the grass begins to grow. Once the grass has grown at least an inch, you are ready to continue the experiment.
  4. Add water to your pitcher, one cup at a time, until it is almost full. Record how many cups you added and then pour the water over the top of your dirt mountain. Record the height of the mountain.
  5. Refill the pitcher with the same amount of water and then pour the water over the top of your grass seed mountain. Record the height of the mountain.
  6. Repeat steps four and five two more times and record your results.
  7. Make observations, draw conclusions, and complete your experiment sheet.

Vocabulary & Memory Work
- ☐ Vocabulary: erosion, weathering
- ☐ Memory Work – Continue to work on memorizing the Types of Rock.

Sketch Assignment: Freeze-thaw Weathering
- Label the Following: water from rain seeps into a small crack in the rock; the water freezes and expands, causing the crack to widen and allowing more water in the next time; the temperature rises and falls, causing the crack to grow until part of the rock breaks off

Writing Assignment
- Reading Assignment: The *Kingfisher Science Encyclopedia* pp. 32-33 Erosion and Weathering, *Freeze-thaw Weathering* pg. 251 in the Appendix
- Additional Research Readings
  - Weathering and Erosion: *DKEOS* pp. 230-231

Dates
- No dates to be entered this week.

Sketch Assignment Week 29

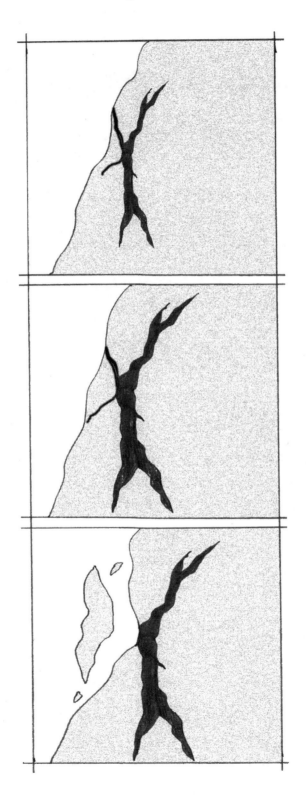

Student Guide Earth Science Unit 5: Geology ~ Week 29 Erosion and Weathering

Do plants help to prevent erosion?

Introduction

Soil is the substance that covers most of the land surface of the Earth. It is composed of rocks, minerals, dead organic matter, tiny living organisms, gases, and water. It is important to life on Earth because it provides the environment and food in which plants can grow. Water and wind can cause the soil to be washed or blown away, which is known as erosion. In this experiment, you are going to see if plants can help to prevent soil erosion.

Hypothesis

Do plants help to prevent erosion?   yes   no

Materials

_____   _____

_____   _____

_____   _____

_____   _____

Procedure

_____

_____

_____

_____

_____

_____

_____

_____

_____

Observations and Results

	Mountain of Dirt	**Mountain with Grass**
Initial Height		
After 1 Pitcher of Water		
After 2 Pitchers of Water		
After 3 Pitchers of Water		

Conclusion

Written Assignment Week 29

Discussion Questions
1. Describe three ways that mineral deposits can be formed.
1. What cycle drives erosion and weathering?
2. How does water break down rock?
3. How does the wind break down rock?
4. What are several effects of erosion?
5. Explain freeze-thaw weathering.

Written Assignment Week 29

Earth Science Unit 6
Weather

Earth Science Unit 6: Weather
Vocabulary Sheet

Define the following terms as they are assigned on your Student Assignment Sheet.

1. Atmosphere – _____
   _____
   _____

2. Atmospheric pressure – _____
   _____
   _____

3. Currents – _____
   _____
   _____

4. Climate – _____
   _____
   _____

5. Coriolis effect – _____
   _____
   _____

6. Weather – _____
   _____
   _____

7. Cloud – _____
   _____
   _____

8. Precipitation – _____
   _____
   _____

9. Supercell – _____
_____
_____

10. Isobars – _____
_____
_____

11. Meteorologist – _____
_____
_____

Student Assignment Sheet Week 30
Atmosphere

Experiment: Does hot water act differently than cold water?
- Materials
  - ✓ 3 Glasses
  - ✓ Red and blue food coloring
  - ✓ Hot water (*Do not handle the hot water without the proper protection.*)
  - ✓ Ice Cold water
- Procedure
  1. Read the introduction to this experiment and answer the question.
  2. Fill one of the glasses two thirds of the way with cold water and add three drops of blue food coloring. Observe what happens and record the time it takes for the food coloring to completely mix.
  3. Next, fill the other glass two thirds of the way with hot water and add two drops of red food coloring. Observe what happens and record the time it takes for the food coloring to completely mix.
  4. Then, pour one quarter of the hot red water very slowly into the cold blue water, observe what happens, and record it on your sheet. (***NOTE*** – *For this part to work, you must pour very slowly down the side of the glass.*)
  5. Check the glass after 30 minutes; observe and record what has happened.
  6. Draw conclusions and complete your experiment sheet.

Vocabulary & Memory Work
- ☐ Vocabulary: atmosphere, currents, atmospheric pressure
- ☐ Memory Work – Begin to work on memorizing the Layers of the Atmosphere. (*See the sketch labels below for a complete listing.*)

Sketch Assignment: The Atmospheric Structure
- Label the Following: Sea Level, Height (be sure to create & label your scale in either km or mi), Thermosphere-outer layer of the atmosphere, Mesosphere-meteors generally burn up as they reach this layer of the atmosphere, Stratosphere-planes fly in this layer of the atmosphere, Troposphere-layer of the atmosphere where weather is created

Writing Assignment
- Reading Assignment: *The Kingfisher Science Encyclopedia* pp. 10-11 Earth's Atmosphere
- Additional Research Readings
  - The Atmosphere: *USE* pp. 134-135
  - Atmosphere: *DKEOS* pp. 248-249

Dates
- 1643 – Evangelista Torricelli invents the barometer, an instrument that measures atmospheric pressure.

Sketch Assignment Week 30

Student Guide Earth Science Unit 6: Weather ~ Week 30 Atmosphere

Does hot water act differently than cold water?

Introduction

      The Sun is responsible for providing light and heat to our Earth. As the Earth heats up, the air and water molecules around the globe also heat up, causing them to rise. As they rise, they cool and fall again. This cycle of rising and falling creates the air and water currents that can be found around the globe. In this experiment, you are going to compare how hot water and cold water molecules behave.

Hypothesis

      Does hot water act differently than cold water?   Yes   No

Materials

_____   _____

_____   _____

_____   _____

_____   _____

Procedure

_____

_____

_____

_____

_____

_____

_____

_____

_____

_____

Observations and Results

	Time (in seconds) for the food coloring to completely mix in
Cold water	
Hot water	

The glass after adding hot water...

Conclusion

Written Assignment Week 30

Discussion Questions
1. What does the atmosphere do?
2. Which gas is known as a greenhouse gas and why?
3. What are the layers of the atmosphere from top to bottom?
4. What role do animals and plants play in maintaining our atmosphere?

Written Assignment Week 30

Student Assignment Sheet Week 31
Climates

Experiment: Do different types of surfaces affect the temperature of a region?
  Materials
  - ✓ 3 Foil muffin cups
  - ✓ Darkly colored soil (i.e., potting soil or soil from outside)
  - ✓ Sand
  - ✓ Water
  - ✓ Light source

  Procedure
  1. Read the introduction to this experiment and answer the question.
  2. Fill one muffin cup three quarters of the way with the soil. Repeat with the sand and water and place all three cups on a surface to sit undisturbed for fifteen minutes.
  3. Touch the surface of each of the cups and rate the temperature from 1 to 10, with 10 being the hottest and 1 being the coldest. Make observations and record your results.
  4. Place all three cups under a desk lamp or out in the sun. Let them sit undisturbed for fifteen minutes.
  5. Touch the surface of each of the cups and rate the temperature from 1 to 10, with 10 being the hottest and 1 being the coldest. Make observations and record your results.
  6. Draw conclusions and complete your experiment sheet.

Vocabulary & Memory Work
  - ☐ Vocabulary: climate
  - ☐ Memory Work – Continue to work on memorizing the Layers of the Atmosphere.

Sketch Assignment: The Water Cycle
  - Label the Following: Clouds become heavy with water and droplets fall as rain, snow, or hail; Water droplets in the air rise and cool, forming clouds; Heat from the sun evaporates water from rivers and seas; Water flows down to the sea in streams and rivers

Writing Assignment
  - Reading Assignment: *The Kingfisher Science Encyclopedia* pp. 36-37 Climates
  - Additional Research Readings
    - Climate: *USE* pp. 194-195
    - Climates: *DKEOS* pp. 244-245
    - Seasons: *DKEOS* pg. 243

Dates
  - c. 5 BC – Strabo, a Greek geographer, proposes the idea of frigid, temperate and tropical climate zones.

Sketch Assignment Week 31

Student Guide Earth Science Unit 6: Weather ~ Week 31 Climates

Do different types of surfaces affect the temperature of a region?

Introduction

A climate is a long-term pattern of weather found in a particular region. Our globe is covered with vastly different climates. These weather patterns determine which types of plants will grow in a region and which type of animals will live in the area. The climate also affects whether people can live in the region. In this experiment, you will be looking at whether a region's surface can help to determine the temperature and climate of the region.

Hypothesis

Do different types of surfaces affect the temperature of a region?

        Yes   No

Materials

Procedure

Observations and Results

Sample	Temperature Rating after Sitting out for 15 Minutes	Temperature Rating after Being under the Light Source for 15 Minutes
Cup #1: Darkly Colored Soil		
Cup #2: Sand		
Cup #3: Water		

Conclusion

Written Assignment Week 31

Discussion Questions

1. What two factors influence the Earth's climate?
2. What is the most important factor in determining an area's climate? Why?
3. What are some factors that can cause a global climate change?
4. What is El Nino?

Written Assignment Week 31

Student Assignment Sheet Week 32
Weather

Experiment: What is the Coriolis effect?
Materials
- ✓ Balloon
- ✓ Two permanent markers of different colors
- ✓ A partner

Procedure
1. Read the introduction to this experiment.
2. Blow up a balloon, draw a straight line around the center for the equator, and hand it to your partner.
3. Have your partner hold the balloon steady as you draw a straight line from the center equator line to the top of the balloon in the same color as you used in step 2. Sketch what the line looked like on the experiment sheet.
4. Now, have your partner spin the balloon in a counter-clockwise direction as you draw a straight line from the center equator line to the top of the balloon in a different color. Sketch what the line looked like on the experiment sheet.
5. Draw conclusions and complete your experiment sheet.

Vocabulary & Memory Work
- ☐ Vocabulary: Coriolis effect, weather
- ☐ Memory Work – Continue to work on memorizing the Layers of Atmosphere.

Sketch Assignment: Seasonal Changes
- Label the Following: Sun's rays, March, January, September, June, the sun's rays are concentrated in the center of the globe - warming both hemispheres evenly, the suns rays are concentrated on the northern hemisphere, the sun's rays are concentrated on the southern hemisphere

Writing Assignment
- Reading Assignment: *The Usborne Science Encyclopedia* pp. 192-193 Weather
- Additional Research Readings
    - Rain and Snow: *KSE* pp. 38-39
    - Weather: *DKEOS* pg. 241
    - Seasons: *DKEOS* pg. 243
    - Air Pressure: *DKEOS* pg. 250

Dates
- 1887 – The biggest recorded snowflakes, which were 15 inches across, fall in Montana.

Sketch Assignment Week 32

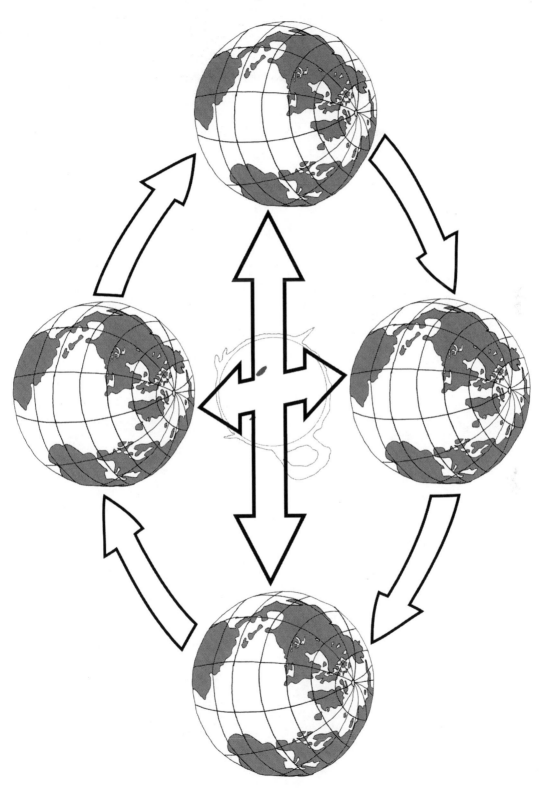

Student Guide Earth Science Unit 6: Weather ~ Week 32 Weather

What is the Corolios effect?

Introduction

The Coriolis effect is a principle that was discovered and explained by a French professor as he was trying to understand the various forces that affect machinery. Although he was not interested in earth science, the force he found also affects the movement of wind and water around the globe, which in turn affects our weather. In today's experiment, you are going to examine what the Coriolis effect is.

Materials

Procedure

Observations and Results

Trial #1	Trial #2

Conclusion

Written Assignment Week 32

Discussion Questions

1. What factors play a part in determining the weather?
2. How does the sun affect the weather?
3. How are areas of low pressure created? High pressure?
4. What are convection currents?
5. Why do we have season on the earth?

Written Assignment Week 32

Student Assignment Sheet Week 33
Clouds

Experiment: How do clouds form?
  Materials
  ✓ 2 Liter bottle
  ✓ Water
  ✓ Matches

> ☹ **CAUTION**
> *Do not use matches without adult assistance.*

  Procedure
  1. Read the introduction to this experiment and answer the question.
  2. Add three cups of warm water to the bottle and cover it with the cap. Wait five minutes and observe what happened.
  3. Squeeze your bottle and release, watching carefully to see what happens. Shake down any excess condensation if necessary so that you can see what is happening.
  4. Remove the cap from the bottle. Have an adult light a match and drop it into the bottle. Replace the cap quickly so that some of the smoke is trapped. Once again, squeeze your bottle and release, watching carefully to see what happens.
  5. Draw conclusions and complete your experiment sheet.

Vocabulary & Memory Work
  ☐ Vocabulary: cloud
  ☐ Memory Work – Begin to work on memorizing the Types of Clouds.

  1. Cirrus
  2. Cirrostratus
  3. Cirrocumulus
  4. Altostratus
  5. Altocumulus
  6. Stratus
  7. Stratocumulus
  8. Nimbostratus
  9. Cumulus
  10. Cumulonimbus

Sketch Assignment: Types of Fog
  📇 Label the Following: Advection fog, fog, warm air, cool land; Frontal fog, warm air mass, fog, cold air mass; Upslope fog, fog, moist air; Radiation fog, land loses heat, fog forms

Writing Assignment
  ✍ Reading Assignment: The *Kingfisher Science Encyclopedia* pp. 40-41 Clouds and Fog
  ✍ Additional Research Readings
    📖 Clouds: *DKEOS* pp. 260-261
    📖 Fog, Mist, and Smog: *DKEOS* pg. 263

Dates
  ⏲ 1803 – Luke Howard, a pharmacist and amateur meteorologist, devises a system to classify the different types of clouds.

Sketch Assignment Week 33

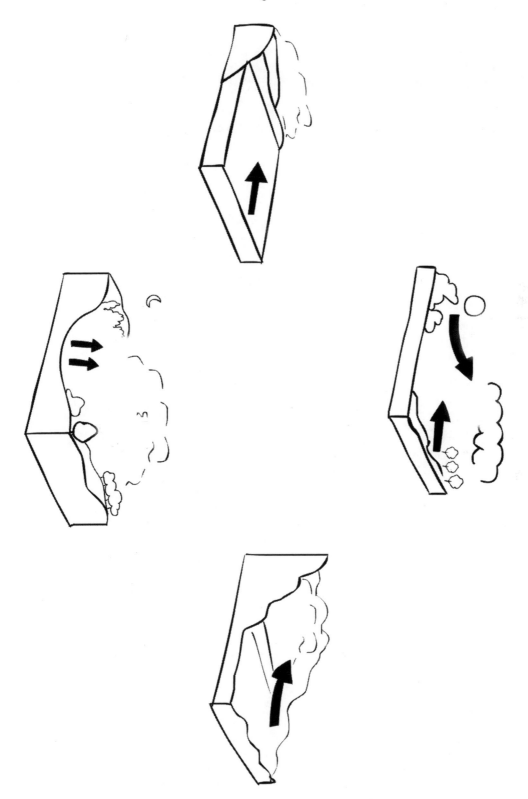

Student Guide Earth Science Unit 6: Weather ~ Week 33 Clouds

How do clouds form?

Introduction

Clouds are mainly composed of tiny droplets of water or ice. The droplet's weight is so insignificant that they float in the atmosphere. Clouds can appear puffy and large or high and wispy. The way a cloud looks depends on how much water it contains and how fast it was formed. In this experiment, you are going to examine how clouds are formed.

Hypothesis

I think that clouds form when _____

_____

_____

Materials

Procedure

Observations and Results

Conclusion

Written Assignment Week 33

Discussion Questions
1. How do clouds form?
2. What factors affect the type of cloud that forms?
3. How are clouds classified?
4. Describe the different types of low-level clouds.
5. Describe the different types of mid-level clouds.
6. What factor is common to all high-level clouds?
7. What is the difference between fog and smog?

Written Assignment Week 33

Student Assignment Sheet Week 34
Extreme Weather

Experiment: How does a thunderstorm form?
  Materials
  - ✓ Large clear round container
  - ✓ Warm water
  - ✓ Red food coloring
  - ✓ Blue ice cubes (*see note below)

  Procedure

  ***NOTE** – You will need to make your blue colored ice cubes the day before. Simply take 1 cup of water, add several drops of blue food coloring, pour into an ice cube tray and freeze overnight.*

  1. Read the introduction to this experiment and answer the question.
  2. Fill your container three quarters of the way full with the warm water. Let it set until the water stops moving.
  3. Gently add several blue ice cubes to one end of the container and several drops of red food coloring to the other end of the container.
  4. Observe what happens and record your results on your experiment sheet.
  5. Draw conclusions and complete your experiment sheet.

Vocabulary & Memory Work
  - ☐ Vocabulary: precipitation, supercell
  - ☐ Memory Work – Continue to work on memorizing the Types of Clouds.

Sketch Assignment: Anatomy of a Hurricane
  - Label the Following: strong spiral winds, dry air sinks, warm moist air rises, low pressure core

Writing Assignment
  - Reading Assignment: *The Kingfisher Science Encyclopedia* pp. 44-45 Wind, Storms, and Floods
  - Additional Research Readings
    - Hurricanes: *DKEOS* pg. 258
    - Tornadoes: *DKEOS* pg. 259

Dates
  - 1971 – The Fujita scale is introduced by Tetsuya Fujita, a professor at the University of Chicago. The Fujita scale is a way of measuring the strength of a tornado based on the damage they cause, F0 being the weakest and F5 being the strongest.
  - December 19, 1898 – President McKinley asks the Weather Bureau, now the National Weather Service, to establish a hurricane warning system.

Sketch Assignment Week 34

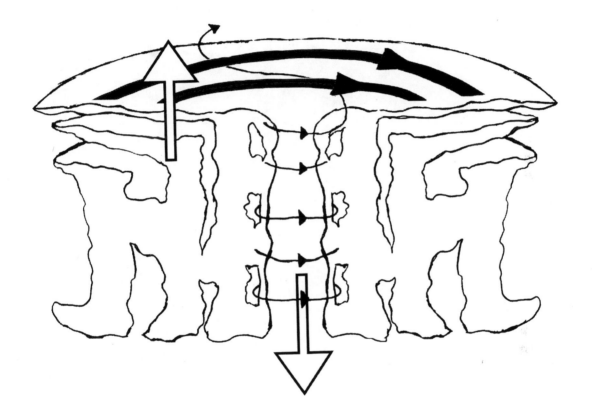

Student Guide Earth Science Unit 6: Weather ~ Week 34 Extreme Weather

How does a thunderstorm form?

Introduction

A thunderstorm is a storm with thunder and lightning as well as high winds and heavy rain. Thunderstorms are produced from cumulonimbus clouds and they typically occur in the spring and summer during the late afternoon or evening. Three main ingredients are necessary for thunderstorm formation. One is moisture, the second is lift, and the third is warm air. In this experiment, you are going to see how two of those key ingredients work together to help to form a thunderstorm.

Hypothesis

I think that thunderstorms form when _____

_____

_____

_____

Materials

_____   _____

_____   _____

_____   _____

_____   _____

Procedure

_____

_____

_____

_____

_____

_____

Observations and Results

___

Conclusion

___

Written Assignment Week 34

Discussion Questions
1. How do winds travel around the globe?
2. How do tornadoes form?
3. What is a tropical cyclone?
4. What is the difference between a tropical cyclone and a hurricane?
5. What can cause a flood?

Written Assignment Week 34

Student Assignment Sheet Week 35
Forecasting

Experiment: What is the Doppler Effect?
  Materials
  - ✓ Battery powered toothbrush
  - ✓ Sound recording device

  Procedure
  1. Read the introduction to this experiment and answer the question.
  2. Place the battery powered toothbrush in front of the microphone of your sound recorder. Turn it on and record the sound for 30 seconds. Then play the recording back and observe how it sounds.
  3. Place the battery powered toothbrush in front of the microphone of your sound recorder again and begin recording. This time move the tooth brush back and forth in front of the microphone several times. Also move the toothbrush far away and close to the microphone. Record the sound for 30 seconds as you continue to move the toothbrush. Then play the recording back and observe how it sounds.
  4. Draw conclusions and complete your experiment sheet.

Vocabulary & Memory Work
  - ☐ Vocabulary: isobars, meteorologist
  - ☐ Memory Work – Continue to work on memorizing the Types of Clouds.

Sketch Assignment: Weather Forecasting Map
  - Label the Following: Isobars, Cyclone, Isobars close together show a sharp change in pressure, Weather symbols

Writing Assignment
  - Reading Assignment: *The Kingfisher Science Encyclopedia* pp. 42-43 Weather Forecasting
  - Additional Research Readings
    - Forecasting: *DKEOS* pp. 270-271
    - Fronts: *DKEOS* pg. 253

Dates
  - 1842 – Austrian physicist, Christian Doppler, proposes that the frequency of a wave changes for an observer as the source of the wave moves closer to or farther away from the observer. The proposal is later tested and found to be true. It is named the Doppler Effect.

Sketch Assignment Week 35

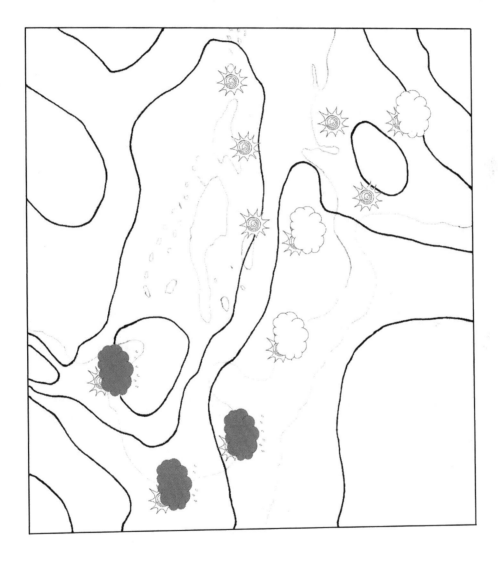

Student Guide Earth Science Unit 6: Weather ~ Week 35 Forecasting

What is the Doppler Effect?

Introduction

Doppler Radar gives forecasters the information they need to predict the severity of a particular thunderstorm. It measures the wind direction and speed using the Doppler Effect. Doppler Radar can also give estimates on rainfall amounts. Meteorologists use this information, along with information from satellites, to provide early warning for severe thunderstorms and accurate weather forecasts. In this experiment, you are going to see how the Doppler Effect works.

Hypothesis

I think that Doppler Effect is _____

Materials

Procedure

Observations and Results

Conclusion

Written Assignment Week 35

Discussion Questions
1. What causes weather systems?
2. How can you use natural clues to predict the weather?
3. How do modern weather forecasters predict the weather?
4. What are some of the tools a meteorologist can use?
5. What is a weather front?

Written Assignment Week 35

Appendix

Astronomy Memory Work

Unit 1

Types of Stars
1. Blue giant – A large, hot star off the main sequence.
2. Red giant – An older star with a cooler outer layer.
3. Neutron star – The tightly-packed collapsed core of a larger star.
4. Main-sequence star – A star plotted in the left-to-right band across the HR diagram.
5. Black hole – A gravitationally dense region of space-time where nothing can escape, not even light.
6. White dwarf – A stellar core remnant of a low to medium mass star.
7. Red dwarf – A cool, small star on the main sequence.

Constellations of the Zodiac
1. Aquarius
2. Aries
3. Cancer
4. Capricorn
5. Gemini
6. Leo
7. Libra
8. Pisces
9. Sagittarius
10. Scorpio
11. Taurus
12. Virgo

Unit 2

Planet Order *(along with the planet's gravity relative to Earth)*
1. Sun
2. Mercury (Gravity: 0.38)
3. Venus (Gravity: 0.90)
4. Earth (Gravity: 1 or 9.8 m/s2) [Moon (Gravity: 0.17)]
5. Mars (Gravity: 0.38)
6. Jupiter (Gravity: 2.34)
7. Saturn (Gravity: 0.93)
8. Uranus (Gravity: 0.90)
9. Neptune (Gravity: 1.13)

Unit 3

Ten Nearest Galaxies and Their Type
1. Milky Way (spiral)
2. Sagittarius (elliptical)
3. Large Magellanic Cloud (irregular)
4. Small Magellanic Cloud (irregular)

5. Ursa Minor (elliptical)
6. Draco (elliptical)
7. Sculptor (elliptical)
8. Carina (elliptical)
9. Sextans (elliptical)
10. Fornax (elliptical)

Earth Science Memory Work

Unit 4

Major Lines of Longitude and Latitude
1. Prime Meridian
2. Equator
3. Tropic of Cancer
4. Tropic of Capricorn

The World's Major Seas and Oceans
1. Arctic Ocean
2. Pacific Ocean
3. Atlantic Ocean
4. Indian Ocean
5. Southern Ocean
6. Mediterranean Sea
7. Red Sea
8. Black Sea
9. Caribbean Sea
10. Gulf of Mexico
11. Hudson Bay
12. Bering Sea
13. Tasman Sea
14. Coral Sea
15. Bay of Bengal
16. Arabian Sea
17. North Sea

The World's Major Seas & Oceans

1. Mediterranean Sea
2. Red Sea
3. Black Sea
4. Caribbean Sea
5. Gulf of Mexico
6. Hudson Bay
7. Bering Sea
8. Tasman Sea
9. Coral Sea
10. Bay of Bengal
11. Arabian Sea
12. North Sea

Unit 5

Types of Rock
1. Metamorphic – Rock that has been changed by heat or pressure
2. Sedimentary – Rock made from particles of sand, mud, and other debris that have settled on the seabed and been squashed down to form hardened rock
3. Igneous – Rock that is formed when magma escapes from inside the Earth, cools, and hardens

The Tectonic Plates
1. Juan de Fuco
2. North American
3. Cocos
4. Caribbean
5. South American
6. Nazca
7. Scotia
8. Antarctic
9. African
10. Arabian
11. Eurasian
12. Indo-Australian
13. Philippine
14. Carolina
15. Bismarc
16. Fiji
17. Pacific

Unit 6

Layers of the Atmosphere
1. Thermosphere – The outer layer of the atmosphere.
2. Mesosphere – Meteors generally burn up as they reach this layer of the atmosphere.
3. Stratosphere – Planes fly in this layer of the atmosphere.
4. Troposphere – The layer of the atmosphere where weather is created.

Types of Clouds
1. Cirrus
2. Cirrostratus
3. Cirrocumulus
4. Altostratus
5. Altocumulus
6. Stratus
7. Stratocumulus
8. Nimbostratus
9. Cumulus
10. Cumulonimbus

The Four Types of Galaxies

Our universe contains four main classes of galaxy shapes that astronomers know about. The first class is the spiral galaxy. These galaxies have a pinwheel shape with a bulge and thin disk in the center. Spiral galaxies are rich with gas and dust. They contain both young and old stars. Many of these galaxies have been named and cataloged because they are the brightest and easiest to spot in the sky.

The second class of galaxy shapes is the elliptical galaxy. They have a round or oval shape with a bulge in the center, but no disk. Elliptical galaxies have a little cool dust and gas. They contain mostly older stars. Astronomers believe that elliptical galaxies are the most numerous in the universe.

The third type of galaxy shape is the barred spiral galaxy. They have a pinwheel shape with a bar of gas, dust and stars running through the center. Barred spiral galaxies are generally rich in gas and dust. They contain both young and old stars. Some astronomers consider the barred spiral to be a subsection of the spiral class, but most consider the barred spiral to be a separate galaxy shape.

The final type of galaxy shape is the irregular galaxy. These galaxies are shaped just like the name implies, with no regular shape. Irregular galaxies are usually rich with gas and dust. They contain both young and old stars. This class contains a hodgepodge of shapes, basically anything that is not spiral or elliptical in shape.

These are the four fundamental classes of galaxy shapes. Astronomers break each class down into further subclasses to help identify various galaxies. They use the shape of the galaxy and how they believe it was formed to classify each galaxy. Even with all the knowledge we have today, astronomers still have a poor understanding of how a galaxy is formed, so shape remains the best way to identify these starry collections.

(taken from The Stars and their Stories: A Book for Young People by Alice Mary Matlock Griffith, pg. 9-15, used with permission)

URSA MAJOR AND URSA MINOR

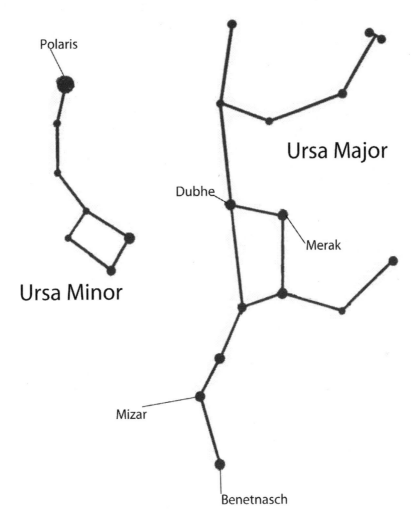

    These two constellations whirl so closely to the pole that in our northern latitudes we do not see them set. Each has a group of stars resembling a dipper; and so they are often called the Big Dipper and the Little Dipper.

    In the ancient pictures of the Great Bear the legs are much longer than shown here; the two hind legs stretch away to the Lesser Lion, and the left front paw touches the two stars that lie just in front of it here; consequently the figure with its long tail and long legs has small likeness to a bear.

(taken from The Stars and their Stories: A Book for Young People by Alice Mary Matlock Griffith, pg. 9-15, used with permission)

CALLISTO AND ARCAS

The Greeks, if we are to judge from their stories, evidently felt that the friendship of the gods was as likely to bring trouble as their enmity. It proved so in the hard experience of Callisto. She was a beautiful maiden whom Jupiter saw and admired. He showed her many a favor, and visited her often; but he endeavored to keep his visits a secret from his wife, Juno, for she was jealous hearted. The goddess, however, learned of the friendship, and she made up her mind to take a severe revenge upon the human maiden.

Callisto had a baby boy, but not even the love of her son could keep her from going a-hunting. While Callisto was eagerly following the chase one day, she had the bad luck to meet Juno. Now, Juno was the most majestic of all the goddesses, as, to be sure, was entirely befitting in the Queen of Olympus. The mere sight of her must have inspired awe in Callisto. And when the poor lady saw the divine eyebrows straighten into the severity of a frown, and the regal, dark eyes, that Homer loves to describe, glow with the fires of jealousy, she must have felt terror as well as awe. Poor, poor creature, her days of happiness were done. With never a pang of pity, the goddess commanded that she change from a woman to a bear. And lo! the thing was done.

Long, long might the babe wait for his mother, and weep because she returned not.

*(taken from The Stars and their Stories: A Book for Young People by Alice Mary Matlock Griffith, pg. 9-15, used with permission)

How was he to know, or his nurse or his grandfather to know, that far in the lonely forest wandered a she-bear, with the frightened heart of a woman in her breast. All his tears could not change that hairy breast back to the white bosom that had pillowed his baby head so tenderly, could not transform rough paws back to the soft hands that had held him so lovingly. And the bear must have suffered, too. For if Juno was unkind enough to wish for revenge at all, she was doubtless unkind enough not to permit the bear to forget that she had been a woman and a mother. I have sometimes wondered if, when, late in the night, the moon shone quietly and sleepily over hillside and town, the bear-mother did not creep up as close to her old home as she dared, and stand looking sadly at house tops and deserted streets, and long for a sight of her little one; only to steal unsatisfied away in the gray, cold dawn, back to her caves and secret recesses. The story does not say, but I think she must have done so. The pain in her heart, I know, was very great, like that in the heart of the father in Matthew Arnold's "The Forsaken Merman," a poem I like to read to my children, and one that you will like, too, if you will ask your teacher to find it and read it to you.

Young people forget quickly. Callisto's little Areas, if he remembered his mother at all, thought of her as dead. He grew through infancy and well into boyhood, active and strong. Like his mother, he became a great hunter.

Then, when he was fifteen years old, came another unlucky day. He was hunting. Suddenly he perceived near him a bear. Neither had heard the other, and both were surprised. At last, he and his mother had met again. It is possible she saw in him a resemblance to what she had been, or to his father, and recognized him. It is certain that he did not perceive his mother in the shaggy form that stood affrighted before him. For a moment they gazed one upon the other. She had no power to make him understand. And he gave her no time to flee away; but, lifting his bow (bravely, as he would have thought, if he had had time to think about it at all), he was about to send an arrow piercing through her heart.

But the friendship of the gods does not bring only evil. Jupiter had not forgotten the maiden he had once loved so kindly. He could not reunite mother and son by restoring Callisto to human form, for one god may not directly undo what another god has done. He could, nevertheless, do what was still better, since eternal fame is better than a long life, even if life be filled with happiness. What he did do, was, first, to change Areas into a bear, too, and then to transfer both mother and son as stars to the sky. There you will see them as the constellations of the Greater and the Lesser Bears.

One would have thought that Juno would rest satisfied, now that the woman she had feared as a rival had become a group of stars, and could not possibly again give her cause for jealousy. But the vindictiveness in her heart was even yet unsated. If her hated rival was to be placed among the never-fading stars, her revenge must likewise be eternal. She sought out her brother, Neptune, god of the sea, told him her story, and begged him, as a favor to her, to refuse to permit the mother and son ever to enter his realm. Neptune granted her request; and consequently the Great Bear and the Little Bear never sink into the ocean, "the baths of all the western stars."

THE TWO BEARS

Change as the stars may from night to ight, these two groups, the Great Bear and the Lesser Bear, can be seen from any place in the northern latitudes at any time of the year. We will commence our search by finding them. They are among the easiest to identify, too. Every boy and girl knows the Big Dipper. If you look to the north, you can hardly fail to see it the first thing—seven bright stars set in the shape of a big tin dipper. Four stars make the bowl, and three the handle. Six of these stars are of the second magnitude and one of the third. In the olden times, the Big Dipper was called by some people David's Chariot, the four stars in the bowl representing the wheels, and the three others the horses. Still other people have named them the Seven Oxen, and others again, the Plowshare. In England, Charles's Wain is the popular name. But the name most commonly used is the Great Bear, the stars in the bowl being in the bear's body, and those in the handle being in his tail, as you may see from the chart.

One of the stars in the Great Bear is quite interesting, because it offers an opportunity to see whether you have good, strong eyes or not. The middle star of the three in the tail is named Iizar, and it has a very small companion, called Alcor, "the Test," because it is used as a test of vision, since only a good eye can see it without the help of a telescope.

The two stars that form the outer edge of the bowl of the Big Dipper are called the "Pointers," because they point to the North Star. When we have found the North Star, we have also found the Lesser Bear, because the North Star is the most important star in this group. The Lesser Bear is like the Great Bear in shape, but is smaller. It is sometimes called the Little Dipper. The North Star is the end star of the Lesser Bear's tail, and the end of the dipper handle; and the four stars of the bowl are in his body. The North Star itself is frequently called Polaris, because a line passed through the earth's poles, and extended into the sky, would come very near it.

Within the last few years it has been discovered that Polaris is whirling very rapidly around a dark companion, and that both together are coming towards us at a tremendous speed.

With a knowledge of where the Little and Big Dipper are, and by the aid of the charts, you can easily find all the other constellations in their due times and seasons.

Freeze-Thaw Weathering

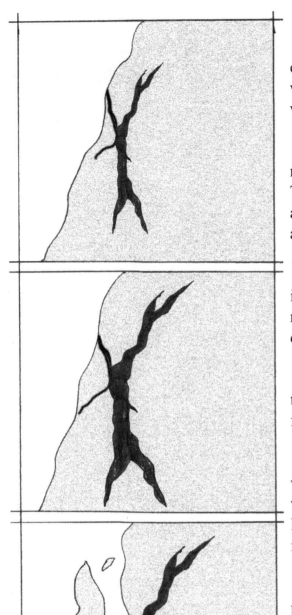

Exposed rocks are constantly being changed by environmental factors, such as wind, water, and dissolved chemicals. This is known as weathering.

One of the key ways that water weathers rocks is through the freeze-thaw weathering. This process of physical weathering takes quite a bit of time to complete, but the effects can be astounding.

In the first part of this process, water seeps into tiny cracks and fractures found in face of the rock. This water can come from snowmelt, rain, or dew.

Then, the temperature drops, causing the water inside the rock to freeze. As the water freezes, it expands, widening the crack.

When the temperature rises again, the water melts and allows more water to seep into the widening crack. Eventually, the crack grows so large that when the water freezes for a final time, it forces a piece of the rock face to break off.

These pieces can be as small as tiny pebbles or as large as boulders, depending upon the size and depth of the cracks.

Activity Log

Activity	Date

What I did/saw/learned

Activity	Date

What I did/saw/learned

Activity	Date

What I did/saw/learned

Activity Log

Activity	Date

What I did/saw/learned

Activity	Date

What I did/saw/learned

Activity	Date

What I did/saw/learned

Activity Log

Activity	Date

What I did/saw/learned

Activity	Date

What I did/saw/learned

Activity	Date

What I did/saw/learned

Activity Log

Activity	Date

What I did/saw/learned

Activity	Date

What I did/saw/learned

Activity	Date

What I did/saw/learned

Glossary

Glossary of Terms

A

- **Artificial satellite** – A man-made object that orbits a planet and is used to gather or relay information.
- **Asteroid** – Large chunks of rock and metal that orbit the sun in the region between Mars and Jupiter.
- **Astronomer** – A scientist who studies the universe and the objects found in it.
- **Atmosphere** – A blanket of gases that surround the Earth.
- **Atmospheric pressure** – The force that the air around us exerts.
- **Axis** – An imaginary line through the center of a planet, around which it rotates.

B

- **Biome** – A region of the Earth that contains unique plants and animals and is characterized by a distinct climate.
- **Black hole** – Formed by a collapsed star, has a very strong gravitational pull.

C

- **Cartographer** – A person who makes maps.
- **Climate** – The long-term or typical pattern of weather in a particular area.
- **Cloud** – A collection of water droplets and dust particles that is caused by a drop in pressure and is visible from Earth.
- **Cluster** – A group of galaxies that are found close together.
- **Coast** – The stretch of land that meets the sea.
- **Comet** – A chunk of frozen gas and dirt that has an orbit.
- **Constellation** – The pattern that a group of stars seems to make in the sky.
- **Continent** – Any of the seven large landmasses into which the Earth is divided.
- **Coriolis effect** – The effect of the spinning of the Earth, which forces winds and currents into a spiral.
- **Crater** – The hole formed in a planet's surface by the impact of a meteorite.
- **Currents** – Patterns of circulation of air or water around the Earth that are caused by the Sun's heat.

D

- **Delta** – A fan-shaped system of streams that is created when a river splits into many smaller branches before it enters the sea.
- **Deposition** – The dropping or leaving behind of rock and other debris by a glacier.
- **Dwarf planet** – A minor planet in the solar system.

E

- **Eclipse** – When one celestial body casts a shadow on another celestial body.
- **Erosion** – The movement of particles of soil, sand, or rock by wind or water to a new location.
- **Erratics** – Large boulders that have been deposited away from their source by a glacier.
- **Estuary** – A wide channel that forms where a river joins the sea.

F

- **Faults** – Cracks in the Earth caused by the movement of its plates.
- **Fossil** – The impression or remains of an ancient plant or animal that has been hardened and preserved in a rock.

G

- **Galaxy** – A body held together by gravity that is made of millions of stars, gas, and dust.
- **Galilean moons** – Jupiter's 4 largest moons, Io, Ganymede, Callisto, and Europa; they are named after Galileo, who discovered them.
- **Gas giant** – A large planet that is mostly composed of gas.
- **Gem** – A mineral or organic stone that is chosen to be cut, polished, and used, typically in jewelry.
- **Glacier** – A mass of ice that gathers at the top of a land mass and slowly flows downhill.
- **Greenhouse gas** – Any gas that traps heat from the Sun, such as carbon dioxide.

H

I

- **Isobars** – Lines on a weather forecasting map that show atmospheric pressure.

J

K

L

- **Lava** – Molten rock at and on the surface of the Earth.
- **Lines of Latitude (parallels)** – Lines that run parallel around the globe, dividing the globe into flat slices; they never meet.
- **Lines of Longitude (meridians)** – Lines that run from the North to the South pole, dividing the globe into segments; they all meet at the poles.

M

- **Magma** – Molten rock inside the Earth.
- **Mantle** – The mostly solid part of the Earth that lies between its crust and its core.
- **Mercalli scale** – A scale that rates earthquakes from I to XII based on the effects of the shaking and the damage caused.
- **Meteor** – A meteoroid that starts to burn as it enters the atmosphere.
- **Meteoroid** – A small piece of space debris.
- **Meteorologist** – A scientist who studies weather and weather forecasting.
- **Moon** – A celestial body in orbit around a planet.
- **Moraine** – Rock, clay, sand, and other debris left by a glacier at the valley floor.
- **Mountain range** – A chain, or line, of mountains connected together.

N

- **Natural cycle** – The exchanging of essential elements, such as nitrogen, carbon, and oxygen.
- **Natural satellite** – A natural object that orbits a planet, such as a moon.
- **Nebulae** – Clouds of dust and gas found in space.

O

- **Oceanic ridge** – A raised ridge on the seabed caused by the movement of the Earth's crustal plates.
- **Oceanic trench** – A deep trench in the seabed formed when one place of the Earth's crust moves under another.
- **Orbit** – The path of one celestial body around another.

- **Ore** – A mineral from which we can extract a useful substance, such as a metal.

P

- **Photosphere** – The surface of the Sun.
- **Planet** – A large globe composed of rock, liquid or gas that revolves around a star.
- **Planetarium** – A device used to project images of the stars or depict the solar system.
- **Precipitation** – Water that falls to the Earth's surface, otherwise known as rain.
- **Prominence** – Large eruptions or loops of flaming gas from the Sun's surface.

Q

R

- **Radio telescope** – A telescope that detects radio waves from objects in space.
- **Reflecting telescope** – A telescope that gathers light with a concave mirror.
- **Refracting telescope** – A telescope that gathers light with a combination of lenses.
- **Richter scale** – A scale that rates earthquakes from 1 to 10, based on the power of the vibrations that travel through the ground when the earthquake occurs.
- **Rocket** – A device capable of delivering objects into space, carries a small amount of cargo and lots of fuel.

S

- **Seismic wave** – An underground shockwave that travels outward from the focus of an earthquake.
- **Solar wind** – A constant stream of particles that flow into space from the Sun.
- **Source** – The beginning of a river.
- **Space probe** – An unmanned space craft that collects information about objects in space.
- **Space shuttle** – A reusable device capable of delivering objects into space, consists of an orbiter, a fuel tank and two booster rockets.
- **Spreading ridge** – A ridge under the ocean that spreads out sideways as magma wells up along its center.
- **Star** – A massive, hot, shining ball of gas.
- **Strata** – The layers of rock found in the Earth's crust.
- **Subduction zone** – An area where two plates collide and one slips under the other, forming volcanoes & deep trenches.

- **Sunspot** – Dark, cooler areas of the Sun's photosphere.
- **Supercell** – A strong, long-lasting, and organized thunderstorm feed by a consistently rotating updraft known as a mesocyclone.
- **Supercluster** – The largest structure in the universe, composed of many galaxy clusters

T

- **Telescope** – A device used to view objects in space.
- **Tides** – The daily movements of the sea up and down the shore; they are caused by the gravitational pull of the moon.

U

- **Universe** – The collection of all the matter, space, and energy that exists, also known as the cosmos.

V

W

- **Weather** – The way the Earth's atmosphere behaves and changes day by day.
- **Weathering** – The gradual wearing down of rock by wind or water.

X

Y

Z

- **Zodiac** – The twelve constellations through which the sun, moon and planets appear to move.

Made in the USA
Columbia, SC
14 February 2025